Death as told by a Sapiens to a Neanderthal

Juan José Millás is considered one of the most important voices in contemporary Spanish literature, and is a prolific bestselling novelist and short-story writer. He is the winner of the Premio Nadal, the Premio Nacional, and the Premio Planeta. As a journalist, he is a multi-award-winning regular contributor to both *El País* and the Prensa Iberica newspaper group. His work has been translated into more than twenty languages, and includes the novels *From the Shadows* and *None Shall Sleep*.

Juan Luis Arsuaga is a professor of paleontology at the Complutense University of Madrid and the director of the Human Evolution and Behaviour Institute. He is a member of the American National Academy of Sciences and of the Musée de l'Homme of Paris, a visiting professor at University College London, and a co-director of excavations at the Sierra de Atapuerca World Heritage site. He is a regular contributor to *Nature*, *Science*, and the *American Journal of Physical Anthropology*, is the editor of the *Journal of Human Evolution*, and lectures at the universities of London, Cambridge, Berkeley, New York, Tel Aviv, and Zurich, among others. The recipient of many national and international awards, he is the author of more than a dozen works.

Death as told by a Sapiens to a Neanderthal

Juan José Millás &
Juan Luis Arsuaga

translated by
Thomas Bunstead & Daniel Hahn

SCRIBE

Melbourne | London | Minneapolis

Scribe Publications
18–20 Edward St, Brunswick, Victoria 3056, Australia
2 John St, Clerkenwell, London, WC1N 2ES, United Kingdom
3754 Pleasant Ave, Suite 100, Minneapolis, Minnesota 55409, USA

First published in Spain by Alfaguara as *La muerte contada por un
sapiens a un neandertal* 2022

Published by Scribe 2024

Typeset in Adobe Caslon Pro by the publishers

Printed and bound in the UK by CPI Group (UK) Ltd, Croydon
CR0 4YY

Scribe is committed to the sustainable use of natural resources and
the use of paper products made responsibly from those resources.

978 1 922585 93 6 (Australian edition)
978 1 914484 85 8 (UK edition)
978 1 957363 75 2 (US edition)
978 1 761385 44 5 (ebook)

Catalogue records for this book are available from the
National Library of Australia and the British Library.

scribepublications.com.au
scribepublications.co.uk
scribepublications.com

Contents

PROLOGUE

Carpe diem

Juan Luis Arsuaga and I were onto the main course of a dinner together, when he asked if I'd like to know how long I had left to live.

"First tell me how much wine we've got left," I said, since the ice bucket was on his side of the table.

The palaeontologist held up the bottle.

"Not much," he said. "We'll have to get another."

"Okay, go on, then," I said, emboldened by the alcohol.

It was early October, and still warm. We were in Seville, where we'd gone to promote our previous book, *Life as told by a Sapiens to a Neanderthal*, and our publishers had put us up in a very central hotel from whose terrace, where we were now having dinner, you could see the remarkable outline of the cathedral and the Giralda bell tower, all lavishly illuminated. The invisible architecture of the breeze, which was a little damp, completed the scene.

The palaeontologist pulled out his phone and opened an app from which, having entered four of five pieces of information about me, he read out that I had twelve years and three months left to live.

"Give or take," he added with a wry smile.

"Give or take," I nodded, doing the maths. "So I've got just enough time to write a couple of novels, as well as the book we've just started this very moment. Thanks very much for the information."

"Don't mention it. It could be a bit more or less. That's the average for Spanish males your age."

"So it's possible we might not even finish this book."

"It's possible. We ought to get a move on," he said, bringing a piece of white meat to his mouth from the sea bass we were sharing.

Then, after a bit of complaining about the illuminatory excesses perpetrated on the monuments of the city — which can be attributed, according to him, to the *horror vacui* of the Spanish temperament — he added: "Since I've got the app open, would you also like to know what you're going to die *of*?"

"I'm not sure," I said. "This sea bass is perfection."

"Well," he went on, ignoring my doubts, "top of the list is heart attacks; cancer's next. Up to the age of seventy, cardiovascular problems and tumours are neck and neck as the main killers, but after that, it's the cardiovascular that surges into the lead."

"And then?"

"In third place, there's respiratory problems, which get grouped together under COPD, an acronym you've doubtless heard about — 'chronic obstructive pulmonary disease'. The next most common killers are way down. Which is basically to say, at your age, people die of being old."

"Well," I said, gesturing for a refill. "Twelve years and three months, you can squeeze a fair bit out of that, if you try."

"But there's bad news for any of us who make it to eighty-five."

"And what's that?"

"Half of them — or half of us — will suffer from dementia of some sort, or will already be suffering from it by then. *Carpe diem*, my friend."

"Since when are we friends?"

"It's a turn of phrase."

"Just in case, I want to be quite clear: we aren't friends. Dessert?"

"Maybe something sweet, with a sherry. Let's see what they've got."

I looked at the cathedral's flying buttresses, the tips and tops of the Giralda. The monuments represented some sixteen or seventeen centuries of existence: a speck of dust in the evolution of the universe. Which meant that my own would not be so much as a blink in the history of the world, nor in the history of mankind and its works. In two novels' time, or just one if death or dementia had their say, I would be a kilo of ash inside a marble urn (I'm assuming cremation, though I haven't made any arrangements).

The palaeontologist must have read my nostalgic look as a longing for eternity, and he attacked the dessert — an exquisite flat sponge cake called a "mostachón" — with a positively childlike expression of gluttony.

"When we get back to Madrid," he declared, brandishing his spoon in the air, "I'll show you eternity. And I don't think you're going to like it."

ONE

The immortals

He was right: I didn't like it.

Eternity went by the name of "naked mole-rat", and what it was, in fact, was a kind of thin rodent, about a foot long, which lived in underground galleries and whose total absence of fur looked like the result of an aggressive bout of chemotherapy, though I soon learned that the animal was immune to cancer, along with all other illnesses. Its skin, which looked very delicate, went from the rosy pink of a newborn hamster to the dark brown of an acorn. It had two disproportionately large, moveable incisors — veritable shovels that took up half its face and gave it the look, if not of a *total* cretin, then at least of being a little slow on the uptake.

As we were saying, it moved about inside these underground galleries, arranged rather like the ones you find inside ants' nests, and which we could see through a longitudinal slice in the earth with a transparent pane on one side (of acrylic or glass, I'm not sure), which in turn gave the habitat the air of a shop window with the animals nervously scurrying to and fro like creatures trying to find their place in the world. I noticed that they did have eyes, though they kept them closed. I asked if the eyes were vestigial, because I really like using that word. *Vestigial.*

"They can see, but living in the dark means they rely more on touch and smell," Arsuaga replied.

The amazing part is that we, the visitors, should also find ourselves inside a narrow tunnel, a gloomy place with an uneven floor, like the one that was the object of our curiosity. The tunnel in question is in Faunia, the Madrid zoo, the part known as "Underground Mysteries", dedicated to the universe that exists below the earth. From the rats' perspective, if they could see us — which maybe they could — our behaviour didn't seem all that different from their behaviour, as human children were running and tripping along our dark gallery just like the rodents were in theirs.

"And you're saying this creature is immortal?" I asked Arsuaga.

"It's the closest thing to immortality I can show you. Mice that live in people's houses have a lifespan of three years. Naked mole-rats live to about thirty, so ten times longer, which is really something for a creature their size."

"And is there any relationship between longevity and size?"

"Of course. A fly lives for thirty days, and an elephant can live to be ninety."

"But that still isn't immortal!" I exclaimed, disappointed.

"Imagine someone guaranteeing you'll live for a thousand years, so ten times longer than the rest of your species. Wouldn't your fellow men consider you an immortal? Wouldn't you feel a little bit immortal yourself?"

I thought about it: a thousand years, astonishing, that's more than Methuselah, a Biblical legend. He had a point.

"And what state would I be in when I reached that age?" I asked.

"Well, here's the rub. This animal doesn't suffer from old age, it never gets cancer or indeed any other kind of illness."

"It can only die in an accident?"

"The truth is, if you remove all external causes of death, we could almost say that it is literally immortal."

"It is also exceptionally ugly, though," I pointed out.

At that moment, a rat more elongated than the others, and with a kind of humpback, appeared in the gallery.

"Does that one have scoliosis?" I asked.

"No, no, she's the queen," laughed Agustín López, the park's head of conservation and its biological director, who was accompanying us on our visit. "The hump is a bulge in the vertebrae, which expand and widen to allow the abdominal cavity to grow — that way, she can have more young."

"And do they reproduce as frequently as mice?" I asked.

"They can have three substantial litters a year. The female has twelve teats."

"So you must be constantly having to get rid of their children," I deduced. "Or does being in captivity mean they're less productive?"

"Don't say 'captivity' — it's a 'controlled environment'."

I thought about old people's homes, where our senior citizens live in captivity, and I imagined a sign outside bearing that euphemism: *controlled environment*; but I said nothing. Instead I asked, "And what happens in controlled environments?"

"They self-limit."

"How do they do that?"

"By eating some of their young."

"Now for the best part," Arsuaga jumped in, maybe to make up for the poor impression I was starting to form of naked mole-rats. "They're eusocial — the good of the group comes first."

"Like bees?" I asked, surprised.

"Exactly. Bees and termites are the most eusocial creatures going. They have a caste system, with each individual assigned an activity. There's a queen, there are sterile workers, and there are the reproductive males. The queen is the only female capable of reproducing."

"How does she stop the others doing it?"

"These creatures," Agustín clarified, "roll around in their excrement and their urine so that they can recognise one another by smell. Well, it turns out that the queen gives off, with her urine, a hormone that inhibits the reproductive capacity of the rest of the colony. When the queen dies, there's a fight to see who's going to take her place."

So we had found ourselves before a mammal with a similar social organisation — if not an identical one — to that of ants and bees. I was quite struck by this. Biology, I always thought, belonged to the literary genre of horror — just as theology, according to Borges, belonged to that of fantasy. Incidentally, when I thought of Borges, it brought to mind his short story "The Immortal", and in turn I remembered the scene where the protagonist passes through an underground labyrinth, much like that of the naked mole-rats, but leading to the City of the Immortals, where he discovers that immortality is actually a punishment, given that death is what gives meaning to life.

Two young boys who had run up from the far end of the dark tunnel stopped suddenly to look at two of the naked mole-rats walking in opposite directions along *their* gallery, meaning the one on the right was forced to squeeze itself incredibly small to get across the top of one on the left.

"The one on top is higher up in the hierarchy," explained Agustín.

The children exchanged a look, saying nothing, though with a surprised expression on their faces (seeing is believing, they seemed to say), then they went on running along our tunnel as nimbly as the rats along theirs.

"What's happening in there," said Arsuaga, who had also noticed the kids, "happens exactly the same out here."

"Tunnel and meta-tunnel," I added, thinking of those tales that take place inside other tales identical to the first.

"What's that?" asked Arsuaga.

"Like Russian dolls."

The idea made me a bit claustrophobic, and I noticed that, despite the cold, a couple of beads of sweat were running down the back of my neck towards my collar.

"Here," Agustín said, "we have two types of naked mole-rats. These come from Somalia. The others are from South Africa, though they look very much alike. In their natural state, a colony of three hundred individuals can occupy the area of several football pitches. And they use different levels for their activities: they sleep in some, store food in others, they set aside some spaces for rubbish … Like in an anthill."

"And you were saying they don't get cancer?"

"Oh no!" said Agustín. "Or heart attacks, or cholesterol. To date, nobody has discovered any cause of death that doesn't come from outside. Besides which, they don't age — they don't have any illnesses, anyway, and they can survive eighteen minutes of hypoxia. Note that there isn't a great deal of oxygen to be had inside these galleries. They could live in an atmosphere like the Himalayas with no trouble at all."

"Right," I said, while trying to calculate by the screaming of the children who were scurrying past us like moles whether we were closer to the exit than the entrance. Which way were you supposed to run if the anxiety grew worse: forward or back?

"And they have no notion of pain, either," said the palaeontologist.

"Quite so," Agustín confirmed this. "They flatten themselves down, squeezing into the grooves, to an incredible degree, and if you cut off one of their feet, they don't feel a thing."

"Their legs really are very short," I pointed out, though I don't know whether I was intending to explain the lack of suffering.

"Yes," said Agustín, "they've evolved to move very nimbly along the tunnels. They walk forwards and backwards with just the same facility. This gives them a great advantage when it comes to predators. They're a complete biological rarity."

"An ant-mammal," I said.

"Don't forget," Arsuaga added, "the humans in Aldous Huxley's *Brave New World* — what were they? Ants!"

"Where they are very sensitive," said Agustín, "is on those two overgrown incisors, those huge shovels that can move independently for digging. They're real tunnellers, and they're the first to sense the signs of an earthquake."

The horror story just got worse and worse.

"Let's get out of here," I said.

"Wait," Arsuaga stopped me. "What would you call that kind of organisation of the colony, where there's a clear division of labour? Altruism, cooperation, barter and exchange?"

"I'd like to think it's a kind of altruism," I said.

"I'm not asking what you'd like to think, but what it is. I mean, you might *want* to be oviparous, but you're a placental mammal. In biology, things just are what they are — I don't know when you're going to get that into your head."

"Fine, well then I'd say it's altruism."

"Look, for now I'll give you some broad brushstrokes, we'll get into the detail another day. So, for example, when these rats go to sleep, they crowd together in a big mass, as a way of reducing their shared body surface and maintaining the overall temperature. This kind of cooperation makes all sorts of sense, because it benefits everyone simultaneously. There's another kind of barter and exchange that takes place over a longer stretch of time: I scratch your back, you scratch mine. That kind of favour isn't mutual, and it isn't simultaneous. I mention this because there are certain attitudes that might seem like altruism at first glance, but are something else entirely. As in game theory, altruism means an individual gaining some benefit identical to the hit taken; only when the other gains the thing you've lost, that's true altruism. But if the other owes you a favour, that's not altruism, it's barter and exchange. Do you follow?"

"I do, but blindly, like in this tunnel, because I have no idea where you're taking me."

"We'll get there."

"Fine, but can we go out into the light now?"

"Let's keep going a little further, I want to show you another kind of eternity. Let's see if you like this one more."

Though our eyes had got used to the prevailing gloom, we were walking slowly, in silence, still rather overwhelmed by the weirdness of the biological spectacle we had just witnessed. We passed a young couple who made us press ourselves against the wall to let them by, since the father was pushing a stroller, practically wider than the tunnel, with a two-year-old wriggling about inside. The shovels of the boy's incisors shone out from his rosy face, giving him an unmistakeably mouse-like look. I thought about how those of us in the tunnel as well as those in the meta-tunnel were all made of flesh and bone. We were all made of that strange material called flesh, supported on a framework of bone.

Flesh. Meat. Bone. Suddenly I understood vegetarians.

Luckily, we humans had invented, by way of compensation, metaphysics.

A few metres on, the palaeontologist and Agustín stopped in front of another of the display windows that dotted our route. I saw that it was a tank of water, with some strange sort of creature lying at the bottom, utterly motionless, its eyes on the gallery, on us. Some five or six inches long, with very short legs, the whole thing a milky white and endowed with a sort of tadpole tail, it gave the impression of being only half-finished.

"It looks like a larva," I said.

"It's an axolotl," said Arsuaga.

God, an axolotl, of course! Cortázar's horror story came to mind, with the guy who goes to the aquarium every day to stare at — and be stared back at by — this amphibian with its hypnotic gaze. He spends hours at the tank, trying to understand what's happening between him and the animal, and one day, when he turns to leave the facility, he sees himself leaving through the amphibian's eyes.

"But isn't that a larva?" I asked.

"It is, but the incredible thing is that, in sexual terms, it's able to become an adult without ceasing to be a larva."

"Like a baby," I said, trying to get things straight, "that was able to fuck and to reproduce while still being a nursing infant."

"A baby that fucks, yes," said the palaeontologist. "Can you even imagine? Eternal youth, another variation on eternity. Better than eternal old age, wouldn't you say? Look at its little legs, they're like a fetus's. They live, though they're almost extinct, in the Valley of Mexico, which was a lake once upon a time but then dried out. Nonetheless there are lots of remnants from its time as a lake."

I took a good look at the axolotl, whose eyes, which were intensely black, stood out like two pinpoints in the middle of that off-white flesh, and I was overcome by the same dizziness Cortázar's character had experienced. The creature seemed to be sucking up my identity, absorbing it. Looking at the axolotl was like staring into the abyss. It was frightening.

"We've seen enough," I said, turning away.

"If it did ever develop fully," Arsuaga added, "it would turn into a salamander."

The animal remained motionless, watching us. Its large mouth was a poor imitation of a smile.

"And if you cut off one of its legs," the palaeontologist went on, "it would regenerate, with those same tiny toes without any nails. Regeneration is another kind of immortality. If we have a leg amputated, why don't we grow a new one? I mean, if we get a wound, that heals. We do have regenerative mechanisms, but not a patch on this guy."

"Well, the liver can also regenerate from a piece," I said.

"And bones: they knit themselves back together. But if you cut an ear off, you aren't getting another one. If you were to take away all outside threats from this animal, in terms of scale, it's immortal; it lives to more than fifteen, which for an amphibian is off the charts."

"Right."

"It might interest you to know that this little animal was

studied by a biologist of some significance, Julian Huxley, brother of the novelist Aldous Huxley."

"Author of *Brave New World*."

"The same. Julian Huxley discovered that injecting the axolotl with a thyroid-stimulating hormone produced another hormone, thyroxine, and that was what completed the maturation process and made it turn into a salamander."

"I believe all this because it happens," I said, "but it really is quite unbelievable."

"Julian and Aldous," Arsuaga continued, "were the grandsons of Darwin's most combative proponent: a biologist who earned himself the nickname 'Darwin's Bulldog' for his ferocious defence of the theories. When Darwin grew old and infirm, and couldn't attend a debate, Thomas Henry Huxley was only too happy to stand in."

"The Darwin's Bulldog thing sounds like quite a story."

"In our story, the one you and I are doing together, I feel a little bit like that, like Darwin's Bulldog," said the palaeontologist, ruefully.

"And who would I be, then?"

"You remind me of Peter Kropotkin, a Russian anarchist."

"I know the one. Besides being an anarchist, he was a naturalist and a prince, except I'm not an anarchist or a naturalist or a prince."

"But you have whims that are Kropotkin-like."

"How so?"

"Just a minute ago, you called something 'altruism' that wasn't at all. Anyway, let's move on. You'll have to make do with those brushstrokes for now."

"I call these brushstrokes 'tooth marks'," I concluded. "It's not for nothing you're the bulldog here."

Back in the car, on our way home, Arsuaga told me to take out my notebook and write down what he was about to say. I did as I was told. He said:

"When we zoom out from the atomic level, people have this idea that everything ends with the individual. But that isn't the case. There's still organisation above the level of the individual: there's organisation at group level. And then, when we zoom out from the group, there's the ecosystem, or the place where individuals and groups of different species interact. The ecosystem doesn't change, it's always the same; it's the individual that changes. My ecology professor at university used to say, 'where there's a lot of life, there's a lot of death' — but I'm not actually sure I agree with that: in fact, there's no such thing as death, because the ecosystem continues. *Life is immortal.* Individuals replace individuals, but the system remains. There is no death, only renewal. Biological systems are far more important than individuals."

"Reminds me of an interview an American journalist did with God, in the 1940s, through a medium."

"You don't say!" exclaimed the palaeontologist, ironically.

"Well, I don't believe in God either, but the fact is that when the journalist asked Him why death exists, he got an answer that really was worthy of a superior being."

"What did He say?"

"That They — because God talked about Himself in the plural, as if there were many of Him — when They were creating life, They never thought of death; that death was an invention of mankind's. What you people call 'death', He added, are movements within life. 'Movements within life' — mark that."

"Wonderful, but going back to the subject of eternity — which is what has brought us here on a Sunday morning when we ought to be making paella for the family — I personally find it terrifying that, with all the advances in medicine, it's possible for us to be eternally old: it's a horrible kind of punishment. Now, if it was a matter of being eternally young, I'd have no complaints. So, when someone comes and offers you eternal life, make sure you check the small print. What do you think Ponce de León went looking for in Florida?"

"Immortality."

"Absolutely not. He was looking for the Fountain of Youth, which is completely different. I'm also interested in eternal youth."

The palaeontologist entered a roundabout slightly too fast and had to brake hard to avoid a car coming from the left. I gave him a disapproving look; he gave a cheeky grin. Then, pulling away again, he said, "What I'm going to tell you now, I don't want you to include. Shut the notebook."

I shut the notebook, but opened the memory box.

"Go on."

"Men of our species aren't overly concerned with youthfulness. I feel well enough, I look at myself and I feel great, I don't need to feel like I'm young or attractive. The problem of attractiveness doesn't worry me overly. What does worry me is not being able to get it up. Fortunately, we men have come across a kind of eternal youthfulness in this respect. It's called Viagra."

"But *why* do you worry about not being able to get it up?"

"I don't know. Why don't you ask yourself the same question?"

"But reduction in libido," I reflected, "also offers some peace of mind. It reduces anxiety. Remember what Luis Buñuel said in his memoirs, that one of the best things about old age was the reduction in that appetite."

"Lies, all lies! Look, Millás, a bit of advice — and it's the only one I'll ever give you: don't believe everything people tell you."

"I happen to agree with Buñuel."

"If you say so, but we aren't thinking at the individual level here. This is about the human species in general, and when people talk about youth, they really mean sexual vigour."

"You don't think it's weird that the thing we least control, which is sex, is also the thing so much of our identity is based around?" I asked. "Guys who can fuck are pretty smug, and they don't even know *why* they're fucking. I mean, what does a person really lose when they lose their sexual desire?"

"A lot, actually, according to the theory of the selfish gene."

"I've never thought about my genes while fucking."

"I'll give you a summary, because I'm not sure you've understood: there are lots of different concepts associated with this thing we call 'ageing' — we don't know exactly what it is, but it's expressed physically, through things like our hair falling out and a reduction in our energy levels. But in general, this process of progressive degradation, of the loss of certain faculties, which paves the way for death, is directly linked to the capacity to procreate. In biological terms, an old person is someone who's either infertile or who has a very reduced capacity for procreation. It's as simple as that. And now, I think we're lost."

And we were indeed lost, because we'd been ignoring the sat nav.

"You won't be eating on time today," said Arsuaga.

"That's a bad business, because I do love eating."

"More than fucking?"

"Well, let's see," I replied, "I don't have the sexual desire I did at forty, which I swear is a relief. If the genie of the lamp showed up and asked me to choose between recovering my sexual vigour of age forty and being able to eat and drink whatever I felt like without the heartburn, I'd choose the latter. No question."

"But it's all part of the same package, Millás. The question is, which package do you prefer: the forty-year-old one, or the sixty-five-year-old one?"

"It was you who was placing the emphasis on sex."

"Fucking is a way of getting to know people, don't forget."

"And masturbating is a way of getting to know yourself. *Nosce te ipsum.*"

"And would you have asked the genie of the lamp for the same thing when you were forty as you would now?"

"At forty I had no experience of old age. Now I do. And with that experience I tell the genie of the lamp: leave me where I am now in sexual terms, but let me eat Mexican food, something really spicy, tequila and sangrita included, without having to pay for it with a rough night."

"I much prefer having tenure to working my way up the greasy pole of the university system," replied Arsuaga. "I did my time on that greasy pole, I virtually wrote the book on it. And yet, I'd give up having tenure if I could be thirty again, any day."

"Well, I wouldn't want to be thirty years younger — what for?"

"You're just being contrary. You prefer being contrary to actually learning things."

Before I could reply, the palaeontologist stopped the car

and told me to get out. I thought he was annoyed, but it was just that we'd arrived outside my house.

"See you, Kropotkin," he said.

"See you, Darwin's Bulldog," I said.

That Sunday afternoon, I reread Borges's story about immortality and retrieved these lines: "There is nothing very remarkable about being immortal; with the exception of mankind, all creatures are immortal, for they know nothing of death. What is divine, terrible, and incomprehensible is to know oneself immortal."

TWO

Live fast, die young, leave a good-looking corpse

I look back now with some puzzlement at the conversation Arsuaga and I had in Seville, over a dish of salt-baked sea bass whose texture and taste my senses can still remember. Death, then, was a mere rhetorical question with which we flirted because it looked attractive enough for us to desire it, though still a thing too terrible to desire us. It was a topic of conversation between two apparently sophisticated people, just another subject, like any we could have plucked out of the air that was a good fit for a starry night. I can write the saddest of poems tonight.

We were safe from death.

Barely three months after that dinner, death had been transformed into something real, palpable, near at hand. Something we needed to address; something pressing.

The naked mole-rat and the axolotl continued to float about in my consciousness like a pair of biological remnants. I think the idea of the remnant, joined together with biology, allowed me to see the scale of my fragility, my insignificance, and my old age, too. Until recently, I was ostensibly an old man — I was one officially, if I can put it like that — but the actual person who lived within that old man was middle-aged, a youngster, a guy who took trains or planes four or five times a month, who worked eight or nine hours a day, who ate out, with friends, publishers, or journalistic colleagues, two or three times a week. He played at being old the same way he played at being neurotic: as a weapon of seduction, because people always like neurotics who joke about their neuroses and old people who laugh about their age.

Towards the end of the year, I had to renew my ID card, and I was given one that expired in the year 9999. When I enquired, believing this was some mistake, they told me that once you've turned seventy, they gave you a card to last the rest of your life. I left the administrative office, then, with a card that certified my identity forever, which is the same as certifying it for never. It meant that the state considered me written-off, it considered me dead. I went into a bar and recalled how excited I'd been when I got my very first ID card, at sixteen. I debuted it in a little leather wallet my parents gave me, into which they had put a five-peseta note.

I was somebody! This was evidenced by that piece of plastic bearing my photo, and by that paper fortune that boasted a portrait of Alfonso X the Wise, if I remember correctly.

Now my wallet contained several credit cards and three fifty-euro notes, but I was nobody, because the final ID card of my existence looked a lot like a death certificate.

As a result of this psychological blow (or at least I think this was why), it was only a matter of days before I suffered an episode that would noticeably weaken me. This was a Friday, when I woke up in the morning feeling distinctly odd.

You could say I woke up feeling hazy, gloomy, a little slow. The radio announced the arrival of an area of low-pressure squall on the left temple of the Iberian Peninsula.

I had a public event scheduled for that afternoon, with the movie critic Carlos Boyero, at the National Library. We were to talk about things each of us had read that had impacted us. They weren't paying much, which still annoyed me, as it was yet further evidence of how little the state thinks of its writers. This being the same state that had just given me a death certificate disguised as an ID card.

But as we were saying, I woke up feeling hazy, gloomy, slow, sullen.

The event at the National Library excited me because it had been a while since I'd seen Boyero, whom I like — but at the same time, I couldn't really be bothered. While I brushed my teeth, I went over the titles of the novels that had changed my life. I had a light breakfast and withdrew to my study to work without having showered.

Around noon, after having sent in my day's article, I headed to the bathroom intending to get ready for the afternoon event. After getting the shower to temperature, I stepped into the tub and wetted my hair, before putting on the shampoo, which I scrubbed into a copious lather.

Naturally, I had my eyes closed.

All of a sudden, because I couldn't see anything, I felt a sort of shift away from myself, a sort of distancing that affected my sense of who I was. It was as if the skull I was scrubbing was my father's, the hands doing the scrubbing were also my father's, and I, too, perhaps, was my father. I thought maybe when I opened my eyes, instead of finding myself in my bathroom at home, I would appear in my parents' and that I'd step out into the corridor and encounter my mother transformed into my wife.

Going hot-air ballooning is an adventure, I won't deny it, but it's an altogether risk-free sort of adventure compared to the dangers of galloping, full steam ahead, astride the imagination.

Panic compelled me to open my eyes just to check that my bathroom was indeed my bathroom and that I was still me. Then, calmer now, I stepped out of the shower, got the foam out of my eyes, which were stinging, put on my bathrobe, and stood looking at myself in the mirror, which was partially steamed up, for a few minutes. I was panting as if I'd been running, and the feeling of strangeness, despite everything, was still there.

I walked gingerly into the bedroom, sat down on the bed, and called out to my wife, who fortunately wasn't far away.

"What's up?" she asked when she saw me looking as pale as I'd just seen myself in the mirror.

"I don't know," I said, "something's happening to me, but I don't know what it is."

"Where does it hurt?"

"It doesn't hurt anywhere, but I don't know what day it is."

"It's Friday," she said. "This afternoon you've got an event with Boyero at the National Library."

So it was Friday. I pulled on the thread and remembered who Boyero was. I pulled on the thread a little further and remembered that I'd been up at the crack of dawn to write an article to send to one of the papers I work for. Then the thread broke and I couldn't pull on it anymore.

"I don't know what's happening to me," I insisted, in response to Isabel's worried, questioning face, "but I feel strange."

To be clear, it was as if my soul had shifted in relation to the place it usually occupied in my body. My kinaesthesia was failing me, that sense that allows you to know, for example, where your hands are, even if you aren't looking at them. I looked at my naked feet, down there, and it was like they were somebody else's. In some distress, I popped a sedative under my tongue, one of those pills I always have ready, in my nightstand. In a few minutes, I noticed a retreating — albeit perhaps only strategic — of the panic, and something occurred to me.

"Bring me the blood-pressure monitor," I asked Isabel.

I've had one of those pieces of equipment in my study ever since I was told that high blood pressure, in the United States, is called "the silent killer", because it doesn't show any symptoms. I take it from time to time, at random, and on the whole it's fine because every morning I take a very mild pill to keep it in check.

At this moment in time, it was off the charts.

I took another of those pills, even though I'd already taken one first thing, at breakfast, and I called my doctor. I started explaining the situation, but my wife soon took the phone because I was still confused and I wasn't explaining myself clearly. The doctor said that both the decision about the sedative and the decision about the pill for the blood pressure had been correct and that in half an hour I should take another one.

"If it hasn't gone down," he said, "you'll have to call 112."

The idea of calling 112 and the ambulance alarmed me, so I put another half of a sedative under my tongue and stayed calmly where I was, lying down, with Isabel beside me. Half an hour later, my blood pressure was still high, but coming down. My soul, my mind, or whatever you want to call it, had rediscovered its place within my psychic geography, like a ship's cargo returning to position at the bottom of the hold, having been dislodged by a sea swell. My feeling of remove from myself began to give way. Of course, I cancelled the event that afternoon at the National Library, and when I felt strong enough, called Boyero to apologise.

I spent the rest of the day sitting there, my gaze rather absent, just thinking about the fragility of it all.

Was that a shot across the bows from old age? Had this been a brush with one of death's wingtips?

In the days that followed, recovered now in physical terms, though in low spirits from the shock, I talked to Arsuaga several times about our plans to write a book about old age and death.

"My perspective has changed," I told him. "Now I'm looking at these things from the inside."

"Stupendous," he said, "because I'm looking at them from the outside. To me, they're still just an object of study. We'll have complementary takes on it."

Arsuaga is eight years younger than me, but eight years, at our sort of age, is a lot. Besides, the palaeontologist climbs, skis, does these great long hikes across the Madrid mountains, and takes part annually in the international cross-country race in Atapuerca. And he does not, as far as I know, experience panic attacks, which protects him, psychologically speaking, from sudden emotional increases in pressure and their neurological after-effects. In short, he both is and feels young.

"I'll see the house of old age and death from the inside, and you'll see it from the outside," I added. "You'll have access to the facade, and I'll be in charge of monitoring the state of the plumbing and the boiler. Some houses' appearances can be very deceiving."

"Not to me," concluded the palaeontologist with that certainty of his.

So it goes, on 7 January he called to invite me on a trip to the scrap yard, as he needed a wing mirror for his Nissan Juke, which he's nicknamed "the Jaca" — or Pony. The Jaca, which did us such good service in the previous book, is an ancient thing, one hundred and forty thousand kilometres old, always with some ache or pain; if it's not one thing, it's another. Right now, it had a pain in one of its wing mirrors (eye pain, I thought to myself).

The scrap yard, which turned out to be the largest in Europe, was twenty-five or thirty kilometres from Madrid, on the Toledo road. It was a vast expanse of several acres arranged like an endless car park, where the corpses of cars scrapped by their owners were arranged by brand. You could get to the brand of your car on foot, if you felt like walking, or take a bus that ran the length and breadth of the whole vast cemetery, stopping in its different areas: the Renault section, for example, or Mercedes, or SEAT, or Volkswagen ... You could find every known ethnicity of car so long as you had a map of the territory.

We opted to walk, though the land of dead Nissans was a bit far away.

"And what other ailments does the Jaca have?" I asked the palaeontologist while rearranging my scarf, because the day was a cold one.

"The light on the display works when you first start the ignition, but then it fades, which is annoying, because that's where all the icons for the radio, the phone, and so on are. I took it to a mechanic, who checked it over then gave me one of those looks, like ..."

"Incurable old age," I said.

"One hundred and forty thousand kilometres. It's hardly that much."

"If we counted our age in kilometres instead of years, I'd be practically the same age as your car."

"There's an app now that can tell you how many steps you take in a day, it counts the stairs you go up or down — so you could actually do that calculation."

"What would you say it is in me that's starting to fail?"

"Well," he replied, "you never want to come skiing with me. That's just a fact. And you don't go hiking in the Sierra."

"The life of men and cars goes gradually: the screens stop working, the arteries harden ..."

"Arteries losing elasticity," he confirmed, "a bad business, it's true. Cholesterol build-ups, all that."

"I take one pill a day for cholesterol and another for my blood pressure. Don't you take anything?"

"Not yet. It's the irreversible I'm concerned with here. A wing mirror you can change, a back tooth you can get an implant for, but what's irreversible is irreversible. The worrying thing in my car is not so much the mechanical parts as the electrics. The mechanics can be solved easily, like our teeth, or a hip, but the electrics are more like the nervous system. I don't know ... My car's old because of the many kilometres it's done, not because of its age. There's a subtle distinction here, take note: it's old because it's lived fast. When people decide to buy a second-hand car, the first thing they ask is how many kilometres it's done. Next, whether it's been kept in a garage, and whether the owner is a travelling salesman or a lady who's only used it to go to and from Mass."

"And you and I would be doing better if we'd only used

our bodies to go to and from Mass?"

"I'm not talking about you and me, it's different species. Species that hibernate, you might say they've been kept in the garage. But there are species that live faster than others. This is one hypothesis about the ageing process that has many supporters. It's called 'life rhythm'. If my car, rather than one hundred and forty thousand kilometres, had done only thirty thousand, and had been kept in the garage, it would be new. What appears to determine the ageing process is the *pace* of the life in question. There are some species that, to paraphrase the rock 'n' rollers, live fast, die young, and leave a good-looking corpse."

"What does 'live fast' mean, biologically speaking?"

"It means a higher metabolism, a greater oxygen consumption."

"Oxygen kills?"

"Oxygen kills. A tin of mussels, if it never gets opened, can last five or six years, but from the moment you open it and the oxygen gets in, the mussels start to decompose. Here's something you'll find surprising: if you measure the heart rate of a mouse, which lives for three years, and that of an elephant, which can live to ninety, you'll see that over the course of their lives they have the same number of heartbeats."

"But the mouse has lived faster."

"Right: its energy consumption has been higher. The best example is the shrew, which only lives to a year old: it's the shortest-lived mammal. Creatures that feed on insects have to eat the equivalent of their own body weight every day, more or less. A shrew is constantly either metabolising or oxidising. For some reason this means it dies younger."

"What does that 'for some reason' mean?"

"We're going to leave that question open for now. What we're looking to confirm at this particular moment is the idea that the more kilometres you do, the more you age. My car has gone through more petrol than the one belonging to the lady who only uses hers to go to Mass. What animals do is they oxidise, so they use up glucose. The way cars use petrol. And — keep in mind — the same thing the cells do by oxidising, consuming like that, is genuinely a kind of combustion, not one that produces a flame, but it does produce heat, because it's a slow kind of combustion."

On either side of the road we were heading down towards the Nissan part of the cemetery, men would appear rummaging around inside the entrails of dead cars. There would be one guy who had got himself a cam belt, another with a shock absorber, another with a brake pad, a steering wheel, an alternator, a brush, a windscreen-wiper, a clutch, a dashboard … Some of these men, because it was only men, had brought their own toolboxes with them, for breaking up the corpses to make it easier to get hold of the part they were looking for.

We stopped to talk to one of them. He was a Moroccan man who sent the parts to Morocco, where a brother-in-law of his resold them for a good price. The stripped cars stood there with their doors open or bonnets up. Often, if you looked inside, you'd find dirty tissues, teddy bears, empty bags of sweets, or crumpled cigarette packets. Some of these cars had come from accidents in which the driver or passengers had died. The mats sometimes had dark stains that could easily be dried blood.

"You get some species," I heard Arsuaga saying, pulling me from my reverie, "that live fast and die young: insect-eating species, and rodents in general."

"And flies?"

"Flies hardly have a life at all, thirty days or thereabouts, but please don't get ahead of yourself, let's be methodical. As you can see, old car parts get recycled. Like I said before, the mechanical parts in the body are relatively easy to replace. The problem is, we're made up of physics *and* chemistry. The chemistry is more complicated. Write this down, in case we forget."

"Go on."

"There's a link between recycling and the meaning of life."

"I've made a living will according to which, when I die, parts of my body they think healthy can be reused: my corneas, my liver, whatever, anything they think is still in a fit state."

"Have you donated your body to science?"

"Not my whole body, because snatching it away from the family, who'd want to hold a vigil over it and stuff like that, seems a bit harsh. But if someone can make use of a valve, let 'em."

"What I was saying about recycling: this is why there's a good analogy between these cars and our bodies. Some part of your entrails will doubtless still be in a good state when you go. And here, remember Ford's law, the carmaker: there's no point in one part lasting longer than all the others. Why design a carburettor, say, that lasts twenty years for a car that's only going to run for ten?"

"But the body doesn't go to ruin all at once. Look how well my hand works, and it's the hand of an older person."

"That's great; most people your age would have osteo-arthritis."

"Sorry to ruin your day."

"Modern medicine in large part takes the same view as mechanics. Got a heart problem? Here, have a bypass. They replace the eye lens when they operate on your cataracts, they give you a liver transplant … they fix this, they fix that, whatever's not working. And now you're going to live to a hundred."

"Like the cars in Cuba, which are mostly from the 1960s or 70s. All from changing the parts one at a time."

"But then you get the zealots, the ones claiming old age can be reversed. That's a whole other concern."

"It's *our* concern."

"No," said Arsuaga, emphatically, "what we're concerned with is the ageing process."

"Incidentally, how's your prostate?" I asked, apparently apropos of nothing.

"As you'd expect for my age," he said so as not to answer.

At this point, practically without realising it, we had reached the dead Nissan zone. There were dozens or hundreds, maybe thousands, I couldn't tell you, in every possible model. Arsuaga looked for one of his, of which there were lots, it being a middle-class car, and he found one with a left-hand wing mirror in good condition. Then he pulled from his pocket one of those multipurpose tools you can use as pliers, screwdriver, knife, file, et cetera, and with a surgeon's skill he managed to prise the mirror from the body of the car, which didn't so much as whimper in complaint.

"You've done that before," I said.

"Of course — why do you think my cars last so long?"

"And what do you do with the mirror now?"

"Now we go over to that building behind you, I take it to the till, they say how much it's worth, I pay, and off we go. But before that, I'm going to show you the machine they use for turning the chassis into those blocks you'll have seen in the movies. I think it's over there — you see that crane?"

So we did indeed head that way, though the cold was intensifying.

"It's started snowing," I said, hoping this might make him give up.

"All the better, it makes it more beautiful. Snow is manna."

The snow was falling on the dead bodies, which within hours it would have covered like an even shroud. We picked up the pace and soon arrived where that curious operation occurred. There was a real mountain of chassis that had been stripped of anything recyclable. They had been left as bones, as bodywork, as pure skeleton. A crane with three huge fingers approached the mountain, picked up one of the corpses, moved a few metres, and dropped it into a large crate, where it was subjected to a triple- or quadruple-crushing that transformed it into a metal block resembling a cube of chicken stock. The same fingers that had dropped it into the crate now lifted it out again and put it down onto a pile with other cubes, which formed a multicoloured wall, a sculpture, an installation, maybe a performance piece — beats me — something one might look at through an artistic lens.

"It's exactly like what they do with old chickens," I said to Arsuaga. "That's what they turn them into, a stock cube that you can then dissolve into your paella."

"These cubes are melted down once again, and the resultant metal is recycled."

"It's amazing: the melting mixes up the chassis of the Mercedes and those of the SEATs. There's no difference between rich cars and poor cars there. All crushed together like that, they become equals, the labourers, the rich. Could be that a modest old car like yours might be reincarnated, depending on its karma, as a Jaguar."

"Could be," replied the palaeontologist, smiling ironically, as he turned away to walk to the facility's main office, which housed the till we needed to visit.

It had started to snow more heavily; it was beginning to settle on the roofs of the dead cars. Arsuaga's head was white, as was my hat, which I would shake off from time to time so that it didn't get ruined.

"You're going to catch a cold," I said.

"Because of this?" he gestured towards his hair. "This is magnificent! You don't know what it is to go for a great hike on the Sierra. Snow's fantastic, I love it."

Standing at the till was a queue of men, each of them with a couple of organs they had obtained from the scrap yard. Some were carrying a plastic bag filled with innards that they laid out on the counter. The assistant valued the scrap at a dizzying speed. He charged the palaeontologist eighty euros for the mirror, which we thought a bit steep.

On the way back, the windscreen-wipers working constantly, with the storm showing no sign of abating, Arsuaga referred to the nomenclature that Aristotle used in the *Metaphysics* to explain how practising physicians, those who devoted themselves to the replacing of bone hips with titanium ones, for example, concerned themselves with the "efficient" or proximate causes.

"While we," he added, "are interested in looking into the 'ultimate causes': why does the heart stop working, why does a life last as long as it does, why is it that one species lives longer than another? I'm not saying that the 'efficient causes', the proximate ones, shouldn't be our concern, but our real objective is the ultimate ones."

"Right," I said, looking with some curiosity at the industrial buildings, on whose roofs the snow had begun to form a sheet.

"In biology," he continued, "the final cause is evolution. Things are the way they are because of evolution. What we ought to ask ourselves is why evolution has not, over a time span of four billion years, made us immortal. A lot can be achieved in four billion years."

"Immortal as a species?" I asked.

"No, no, immortal as individuals. What's evolution been up to during these four billion years? Wouldn't it be an advantage for the individual to be immortal?"

"I've already told you, Arsuaga; according to God, death doesn't exist. Death is a movement within life. Calling this movement 'death' implies a lack of a sense of humour."

"Biology hasn't got a sense of humour, Millás. Evolution comes about through a process called 'natural selection' — of those who live longest — because biology learns from the things that go right, not that go wrong. Biology doesn't learn from those who live only a short time. Natural selection picks the individuals with a greater capacity for survival, and those are normally the ones who have more children. It's important to hold onto that: natural selection proceeds from individuals, not from the species."

"If we individuals don't die or don't experience what — with a significant lack of humour — we call 'death', then we'd become a veritable plague."

"And who cares about that?"

"Me. The whole idea makes me uneasy because I identify eternity with a never-ending Sunday afternoon. Since I was little, Sunday afternoons have made me panic."

"Still," said the palaeontologist, indifferent to my unease, "we have to look at what Aristotle called 'ultimate causes', at why mice live for three years and elephants live to ninety. We have to go and ask evolution, since she's the one who knows, to explain it to us."

"And where does this Lady Evolution live?"

"And not only to explain it," he went on, "but to vanquish death, once we have managed to understand it."

"We should hurry, because according to your own calculations, I haven't got all that long left."

"For now, we've learned that a very interesting analogy exists between machines and living beings."

"And that the ageing of a car doesn't depend so much on how old it is as on how many kilometres it's done — that is, the wear and tear."

"Biologists turn to cells to understand old age. It's there, in the cells, that's where the secret lies. The problem would be solved if all the cells would just split infinitely."

"And avoid going wrong," I observed, thinking about cancer cells.

"That is part of the answer. We'll carry on with it, but as a summary — write this down — among the things we've learned today, a mouse covers the same number of kilometres in three years of life as an elephant does in ninety."

"The mouse is a rock star and the elephant more the easy-listening type."

"Right. We should ask the question: why does metabolism age and ultimately kill?"

"We should call it '*murd*abolism'."

Arsuaga pretended not to have heard my joke and just went on doing his own thing: "If we come from a cell that's multiplied to produce such a complex being, how come this being can't be maintained indefinitely, why doesn't it *carry on* getting repaired? That's where we'll find the ultimate cause, which we shouldn't get mixed up with bodywork and paint. We'll leave it there for today."

When the palaeontologist says "We'll leave it there", he's lying. He's never able to leave it "there". And indeed, after a brief comment on the beauty of the snow, he resumed the task.

"I'll take the Epicurean position," he said. "Death is not to be feared, because when you are no longer, it is no longer, and when it happens, you aren't here anymore. Next on the chain, among my heroes, is Lucretius, who sees us as nothing more than atoms randomly moving around. Atoms combine randomly and give rise, with neither logic nor meaning, to the reality we see before us. Lucretius's materialism is fundamental to modern science. Richard Feynman, in the twentieth century, says much the same thing, the difference being that, for him, atoms combine according to certain laws, which are the laws of physical matter."

"Feynman takes chance out of the equation," I added.

"It's a very important nuance. Oxygen does not combine with carbon in any old way to produce CO_2. It's *two* oxygen atoms to *one* of carbon."

"There are laws, then."

"There are laws, which implies an advance in the understanding of physical matter."

"The result, in any case, is that we're made up of atoms in motion."

"Of course. But let me tell you who the final hero is in this chain formed by Aristotle, Epicurus, Democritus, Lucretius, and Feynman."

"Who?"

"Jacques Monod, the author of *Chance and Necessity*, which is a phrase taken from Democritus. In that book, Monod says we're all alone in an indifferent universe that has arisen by chance. Pure Epicureanism. And, in opposition to this school of thought, we have the animist one, which maintains that there's a sense to it all, because nature's *so* wise. To be clear, that's the point made by the likes of Rodríguez de la Fuente. It's a branch of magical thinking. There's a meaning to death because we thereby make a place for those who come afterwards. Every culture has tried to find the meaning of death."

"Of death and of Sunday afternoons," I said, because though it was Thursday, it was exactly that atmosphere, that of a Sunday afternoon.

"Are you writing all this down?" asked the palaeontologist, his eyes never leaving the road, as the driving was getting dangerous.

"More or less," I said.

"Monod was in the Communist Party," he said, "which created a lot of difficulties for him. In those days, Marxism was often said to be the official religion of science, but he didn't agree. To his mind, Marxism was a kind of magical thinking. Marxism holds that history has a direction. According to Marxism, mankind's emergence was inevitable because this directionality develops in a way that's both predictable and logical. According to Monod, Marxism, Judaism, and Christianity are all attempts to cling onto the animist way of thinking."

"I'm not an animist, but I do feel nostalgic for animism."

"Then you're nostalgic for something that never was."

"'There's no nostalgia worse than longing for what's never happened', that's what the Joaquín Sabina song says."

"Being nostalgic for animism," he replied, "just seems like another way of being an animist."

"Whatever you say."

"Certainly, though, this idea that nature is wise, which is such a distinctively animist idea, constitutes a kind of religion, a lay religion, if you like, and it's a consolation to people."

The snow, which previously had fallen copiously but gently, was now hurling itself with fury against the windscreen, thrilling and invigorating Arsuaga. Not me. Me it reminded of the snowfall on a Sunday afternoon in my childhood, an eternal Sunday afternoon that those white flakes, watched from the window of a cold, ramshackle bedroom, made even *more* eternal, if that's possible. My whole life I've thought about that afternoon. I've never managed to escape it.

"Certain strains of animism, certain strains of ecologism, as well as the incredibly famous Gaia hypothesis — which posits the earth as a superorganism — are all ways of clinging onto magical thinking in order to overcome the anguish of life's lack of meaning. But really, Millás, there's nothing to support such hopefulness: we've arisen by chance in an indifferent universe, a universe that isn't even cruel or hostile. It's much worse than that: it's indifferent."

He seemed to be talking about the indifference of the universe with the passion of a preacher, as if adherence to this creed could establish itself as a new kind of religion, but I said nothing so as not to encourage his teacherly impulse, which was boundless. Finally, since he also fell silent, I was forced to say something to lighten the oppressive weight of that white silence: the silence of the snow that collected, thanks to the back and forth of the wipers, on either side of the windscreen.

"But it's possible to create meaning without being an

animist," I said. "You can create meaning from a materialist perspective. Indeed, if the human being is anything, that's what it is: a tireless creator of meaning."

"I'm not so sure. Individually, perhaps, but at a collective level, it's a bit more of a shitshow."

"A society without a purpose, without direction, is a society that's disjointed, broken."

"And that's lovely."

"What is?"

"Belonging to a group, a tribe."

"It's not that it's beautiful or ugly, it's that we're social animals, it's that the group is an individual. Look at what the Paco Ibáñez song says, the lyrics are from a poem by José Agustín Goytisolo: 'A man alone, or a woman, taken individually, are as dust, are nothing, nothing.'"

"Songs are your solution to everything," replied Arsuaga, "but to turn the individual into a group, you've got to turn to animism, something to cling onto … That's where ritual begins."

"The project of an economy growing two per cent next year might be merely materialist, but it's still a project."

"But there's no consolation in it."

"Does that necessarily imply that it doesn't create meaning?"

"I don't think so. But don't get me wrong. It doesn't follow from the fact that everything's meaningless that we have to give up being good. Epicureanism, for instance, was very badly misunderstood — it was taken to be a form of hedonism. And Epicurus himself was a very sober sort. What he was saying is that there's no animism, there's nothing beyond this atomic dance, and that you can't find consolation in belonging to a grand project."

"This atomic dance — forgive me if I harp on, but you always do — shows that death doesn't exist. What's strange is that dead bodies do."

"The individual dies, of course it dies, that's been the case ever since we moved beyond being single-celled organisms."

"It doesn't die, it gets transformed, recycled — you said it yourself. What dies is self-awareness."

"Tell that to the people grieving at a funeral. 'Oh, don't worry, your dad isn't dead, he's still in the ecosystem, he's a part of the biosphere.'"

"Fine, fine, but is it or is it not true that this atomic dance is a set of movements within life?"

"Well, now you're talking like an Epicurean."

At that point, it turned out that we'd arrived, and Arsuaga was inviting me to get out, quite convinced that he was dropping me at my house, while in reality, he was abandoning me in that faraway Sunday afternoon in my distant childhood.

And it was still Thursday.

THREE

Eros and Thanatos

Whenever I mention my ailments to him, the palaeontologist says, "In nature, there's no such thing as old age or decrepitude. In nature, you're either in your prime, or you're dead."

"How so?"

"If a gazelle, which needs to run ninety-five kilometres an hour to get away from its predators, can only go ninety, then it's dead."

"Right."

"And if a young fallow deer breaks a leg, it won't last more than a couple of hours."

"So the decrepitude associated with old age is a product of culture?"

"We'll get onto that."

In order to demonstrate this, he takes me on a trip to the Complutense University of Madrid's Veterinary Faculty, where, as well as hosting surgeries for all types of animals, they have a hospital for doing operations on everything from cats to horses, from rabbits to goats.

"I do dissections on dead animals there," Arsuaga continues as he drives up the M-40 towards University City. "Primates.

We aren't authorised to dissect large mammals, nor do we have the set-up. But what I'd like to do is palpatory anatomy, external anatomy, on the surface of things, like the French do in the medical departments at universities, or like Spanish physiotherapists do. I'd love to come to an understanding of the human body at the university swimming pool, prodding them here and prodding them there, but I'd doubtless be reported, or something would happen to me. Physios don't have that problem, or doctors, but a palaeontologist … I don't know, I've never had the guts to try."

"Speaking of decrepitude," I say, "I went to get some new glasses made because I'd lost my old ones, and the lady at the optician's suggested I go to the ophthalmologist, as it wouldn't be surprising if I had cataracts."

"And did you go?"

"Of course, what choice did I have?"

"And?"

"I don't have cataracts."

"Well, you should."

"That's just what I wanted to tell you: that once you've turned seventy-five, you're required to be old. There's such a huge pressure to be old, if not to die. I went to the dermatologist because my back and scalp were itchy. He said it was an old-age problem, as people my age stop producing fat and our skin dries out. He recommended a soap and some creams. But the symptom persisted. I did a bit of research online, and turns out it's a very common symptom, which affects all ages. The most common cause is that catch-all, 'stress'. Then I came across a guy who says his itching stopped with antihistamines. I start taking the antihistamine and it vanishes. In other words, it wasn't old age. So what do you make of that?"

"I don't know, I really don't, but shall we talk about your ailments another time?" he says while parking the Nissan Juke outside one of the faculty buildings.

It's nine in the morning on an unpleasant day in mid-February.

"The point is, they aren't ailments specifically related to old age," I insist. "An allergy can hit you at any age."

"Right," the palaeontologist concedes.

"I do believe, Arsuaga, that writing this book has fucked up my life. I hadn't realised I was old until we started writing it."

"Sure. Have you written down what I've been telling you?"

"About what?"

"About either being in your prime, or being dead — in nature."

"Yes."

"Well, you'd be dead — in nature."

"Except I seem fine."

"Don't believe everything you tell yourself."

We are received by Iñaki de Gaspar, who introduces himself as a professor of anatomy. He and Arsuaga work together on the master's program in human evolution. On primate anatomy.

"But it's my job to explain to the undergrads the anatomy of the different species we have: dogs, cats, horses, cows, and others. So basically, I'm an anatomist," he concludes.

Almost immediately after him, I get introduced to Lola Pérez Alenza, director of the hospital, the centre attached to the faculty where the veterinary students do their practice.

"We teach them how to operate," she says, "and to hold surgeries for dogs, cats, rabbits, exotic animals …"

"And do you charge a lot?"

"We've got standard rates that have been approved by the university's Social Council, and we really couldn't do without them. Keeping this hospital running costs a huge amount."

"We've come here," Arsuaga says, butting in, "rather than to a human hospital, because we're interested in the ageing process of other domesticated species. I say 'other' because, as I've explained to Millás on various occasions, we are also a domesticated species."

"Self-domesticated," I specify.

"We'll see if the processes around ageing and death are the same in these species as they are in us," Arsuaga continues, oblivious to my added nuance, "and then, after this session with Thanatos, we'll address Eros, and love. This is going to be quite the show, I promise you."

With these words, we venture down a hallway where I'm introduced to Consuelo Serres, dean of the faculty.

"But everybody calls me Cuca," she says.

I've asked the palaeontologist over and over not to introduce me to too many people because it's only a matter of time before I get muddled. I write novels without very many characters because I belong to a family with nine siblings where there were constant identity mix-ups. At school, I was always being called by one of my brothers' names, and even our own parents frequently got it wrong. I once caught my father talking to a friend who was marvelling at how he was able to manage such a large family. "Oh, it's not really that hard," my dad replied, "from the fourth one onwards, you don't even remember their names." As it happened, I was the fourth, so those words left a profound impact on me. You ought not to have more children than you can recognise, nor put more characters into a story than you can handle.

But there was no way the palaeontologist was going to do as I asked, not in this matter or in anything else. He does what he wants. Now, while I'm exchanging a few polite words with the dean, whose name I've forgotten, I hear him saying behind my back, "Since we Spaniards are masochists and we like beating ourselves up, I've explained to Millás that this faculty is one of the ten best in the world."

"Well," Iñaki de Gaspar corrects him, "it's actually number fourteen in the Shanghai Ranking."

"Fourteen's not that bad, either," I point out, trying to be conciliatory.

A silence descends, broken only by our footsteps as we make our way down the corridor. I suddenly realise that being a vet is more complicated than being a doctor, because a doctor only studies one animal species (ours), and within that, they tend to specialise in one organ or system (kidneys, circulatory or digestive systems, et cetera), while a vet, although they too specialise, has got to have some general knowledge applying to all species. Lola Pérez Alenza, for example, is a specialist in endocrinology and in certain types of dog and cat cancers. Cuca (the dean, her name has just come back to me) is a specialist in horse reproduction.

"A kind of horse gynaecologist?" I ask.

"Sort of," she says.

I ask Lola why vets have less social status than people-doctors, given that, in terms of the complexity of what they know, the opposite should be the case.

"That's true for Spain," replies the oncologist, "and I think it's down to a historical question. During Francoism, our profession wasn't very highly valued, because one of the presidents of the republic in exile, the socialist Félix Gordón Ordás, was a vet. Franco used to say: 'Rid yourself of Jews, masons, communists, and vets.' But things are changing — quite considerably, in fact — because pets are now considered family members, and get the same kind of health treatment as children or grandparents. If that wasn't enough, it's the vets who take care of the feeding and health of the animals we eat. The pandemic has highlighted something very positive. What happened in Wuhan couldn't happen here, because all our markets are supervised by vets. It would be impossible to have live animals and dead ones on the same stall. Everything here is very well organised. If you go to the fish part of Mercamadrid, or any other market, you'll see they're totally compartmentalised and well monitored. It's us who ensure that the animals put on sale are healthy. While they're alive, we take care of their wellbeing, creating suitable conditions for them, and reducing the stress involved in putting them to sleep. In China, in these areas, vets don't feature, nor are they expected to. All these things have combined to change Spanish people's perception of our profession."

For a few moments, I just think about my own ailments and wonder if it wouldn't be better to put myself into the hands of a guy who's in veterinary practice instead. Or a woman, since there seem to be more of those.

"What are you thinking about?" Arsuaga jumps in.

"That maybe I should put myself in the hands of a vet."

"You absolutely should," he confirms. Then, returning to his audience, he adds: "Millás is very preoccupied with the subject of old age and Thanatos. Not me — Eros is more my thing."

The long corridor we're walking down is flanked by doors on both sides, like the sort of corridor you'd find in a dream.

Behind each one, there's a consultation underway.

We're allowed to go in to some of these, where we meet the individual vet who's attending and can chat to him or her. The conversations are brief because they usually have an animal on the table undergoing diagnosis. Since the animals don't talk, the information their owner supplies is essential: "She's been a bit depressed," or, "He's aggressive," or, "She's scratching more than usual."

I stop at the surgery of José Luis, a dermatologist whom I tell my story of the itching on my back and scalp, but the man is very sensible and won't venture a diagnosis without examining me. He tells me, in any case, that a lot of cats and dogs come to him with the same problem as mine.

"It's one of the most common illnesses," he adds. "It's usually an allergy to pollen or mites. To anything. Right now, it's cypress season, for example, and arizonicas in general. Their effects tend to be respiratory or cutaneous."

I ask whether allergies are going up in domestic pets as fast as they are in humans, and he says there are no epidemiological studies for animals, but that we might be at about fifteen per cent now, especially in Madrid and the other big cities, because pollution is an aggravating factor.

"Among humans," he explains, "it's calculated that by the middle of the century, fifty per cent of the population will suffer from some kind of allergy."

We keep walking and stop at a surgery where they are doing a neck X-ray on a Doberman, who is fortunately under anaesthetic. On the monitor, we can see just how perfect the vertebrae are. They look like they're made of steel. A bit further on, we find an exotic-animals surgery, where I see a frightened rabbit in a vet's arms.

"This patient is an old one," the specialist tells us. "He's eight and they live an average of between seven and nine. When I finished my degree, they were living between five and seven, but with veterinary care, vaccines, and diets, their lives have been prolonged considerably. We've got patients of up to twelve or fourteen."

"And do the females keep on giving birth right to the very end?" Arsuaga asks.

"The problem with the females," says the vet, "is that about eighty per cent of them, from the age of three and a half on, have uterine tumours. This is due to their fertility, which really is remarkable: female rabbits are on heat almost twice a month. Their uterus regenerates cells incredibly fast their whole lives; this increases the chance that cells will produce tumours. One reason they didn't used to live so long is because they didn't get spayed. We recommend it for all of them now. On a farm, they get put to sleep at two or three. Once the tumour has shown up, their life expectancy tends to be under a year and a half because it metastasises: lung, liver …"

"That's an important point," Arsuaga tells me. "From a certain age, cancers start to develop. This is a subject we need to go into in more detail. Prolonging the life of any species means controlling cellular proliferation in some way or other. There's no way of attaining immortality without controlling the cellular division that, sooner or later, gives rise to tumours."

"And what about the rabbit you're holding, what's the matter here?" I ask, rather overwhelmed by the cellular proliferation and by the look in the eyes of this animal, which seems to be aware of its decrepitude and the nearness of its death.

"His name's Kenny, he's a male, and like I said, he's eight years old. He came in because of a limp — he's had arthroplasty four years ago because of a dislocated hip. Apparently he's had a fall and is limping badly."

"Where does he live?" I ask.

"In an apartment," the vet replies. "Rabbits are very popular pets just now because they're so intelligent, they're interactive, and they're very clean. They can be allowed to run free around the house, and only go into their hutches to eat or do their business. They're more affectionate than cats and don't need a lot of space."

As we walk away, I think about how the frightened Kenny, in nature, would have been food for vermin long ago. The palaeontologist is right: prime of life, or death.

I don't know how it happens, as the rabbit's medical history has rather disturbed me, but somehow we arrive at a cosy little room, about which Lola Pérez Alenza has the following to say:

"This room is for cardio, but we also use it at one very important moment in our work, which is when we need to tell the owner of an animal that the best thing is to put it to sleep due to some terminal prognosis or another, and the fact it's suffering. We help them in their decision by offering both objectivity and tact. It's such a difficult situation because pets are a part of family life, and their absence leaves a gap."

"Right," I say. I'm thinking about my cat, who, according to my calculations, must be in the final third of her life.

"During the actual procedure," Lola goes on, "we dim the lights, we sedate them a little, and then we give an injection that causes immediate heart failure. We don't put music on, but in the US they do. The animals don't suffer at all. We've got to make sure they don't feel any pain. Keeping the *owners* from any pain is impossible, but we can at least soften it by being there with them."

"And what do you do with the body?"

"Some people donate them for students to do practical work. Others get collected by a firm that cremates or buries them, as there are cemeteries specifically for animals."

"The owner can't bury it in their garden?"

"No, that's banned on hygiene grounds. If they die here, they can only leave here with specialist companies."

We exit the cosy little room, the euthanasia room, the lounge of Thanatos, and soon find ourselves in the endoscopy area, where a couple of very young vets (one a young man, the other a young woman) are investigating the large intestine of a ten-year-old dog that's lying asleep on a table. It's a small breed, and very furry, and so lying there like that, so still, it looks like a teddy bear.

Inside the teddy bear, however, thanks to a monitor, it's possible to make out perfectly organic rucks and creases, into the depths of which the endoscopy tube advances with the camera, which has a light to illuminate the mammal's deepest recesses. It really feels like we ourselves are descending into a damp cave.

"What's wrong with him?" I ask.

"He's got a lump in his rectum, we think it's a tumour, but we've got to biopsy it to see if there's something else," she says. "We've only just started. We've reached the anal sphincter, and we've got to get to the end of the large intestine."

While she talks, the young vet is slowly feeding the tube with the exploratory camera into the animal's body. Her colleague, at the monitor, is operating some buttons to turn the lens this way and that in order to detect any malformations that might appear in the intestinal wall. Thanks to a particular diet and some earlier enemas, the dog's pulsating innards are clean, with no traces of food or faeces. The exploration has something of a journey to the centre of the earth about it. Hard to believe that the intestine, barely a few centimetres from the body's surface, is nonetheless in a whole other dimension of reality.

"Now," says the young woman, without looking away from the monitor, "we're in the descending colon. We're going to turn down the transverse colon. And we'll have to take another turning after that. For now, no more suspicious lumps, only the one we found at the entrance to the rectum."

"And how did you find that one?" I asked.

"Blood in the faeces. That's why he's here. Look, we're coming to the end of the large intestine now. We don't go through the valve connecting to the small intestine unless there's a reason to. We can insert forceps or the endoscope, but there's no point on this occasion as the problem's elsewhere."

Lola now leads us to the floor for animals that have been hospitalised, with its complex arrangement of doors and corridors in which I soon lose my bearings. It's like I'm still inside the twists and turns of the dog's intestines. I suddenly feel like I'm nothing more than the bearer of a digestive system that occupies the central part of my body, from mouth to anus.

While we walk, I am told that dogs and cats do not have heart attacks, but pigs do.

"I think it's because of their anatomical configuration," Lola adds. "Dogs and cats also don't have very high cholesterol, even though they do eat meat."

In one of the clinic's cages, we visit a Yorkshire terrier, aged seven, who came in the night before, in an emergency, with convulsions they have already managed to curb. It's suffering from kidney failure, a severe case for its age, as well as hypertension.

"She's in a bad way," concludes the lead vet.

The little terrier seems disorientated and gives us questioning looks. "What will become of me?" she seems to be asking.

I can identify with her disorientation and feelings of defencelessness.

"Dogs' faces," one of them tells me, "are very expressive. Those of cats are much harder to read."

"Will she survive?" I ask.

Their unconvinced looks leave little room for interpretation.

"Her owner knows the prognosis is bad, but he hasn't yet taken 'the decision'," they add. "We're giving him a couple of days to think it through, while she's hospitalised."

We visit more dogs with different pathologies. One of them is suffering from a non-infectious pneumonia because some food has got into its lungs. It's not fatal. In the next-door cage is another that arrived ten days ago with a gastric dilation. At this point, we need to stand a little to one side to make way for some people bringing another dog back from theatre after an operation on an eyelid tumour. He's a large black Labrador. He's still asleep, with his whole tongue lolling out.

"When he gets intubated," somebody explains, "they pull his tongue out like that. If it turns blue, it means he's not breathing properly and we have to do something."

As I begin my retreat from the area — so much decrepitude, sickness, and death is more than I can bear — I'm stopped again to be shown a Spanish water dog who is "good-natured and with suspected pancreatitis". "Suspected pancreatitis," I repeat to myself over and over like a mantra. Sounds good.

While the group continues with its medical visit, I manage to take Lola aside, for her to explain old age in domestic pets.

"It's very different from people," she says, "in the sense that animals adapt to old age better than humans do. They get pains in their knees or their hips, like we do, but they adapt to the changes of age with decorum. My cat, who is sixteen, is the boss of our house. She has osteoarthritis and I give her medication to relieve her pain. She moves around less than she used to, but with great nobility. She's very much the grande dame."

"A human in that situation," I say, thinking of myself, "would have had a fall by now, either that or he'd spend the whole day complaining."

"Exactly," Lola nods. "He'd spend the whole day asking why this had to happen to him. Animals don't lose their composure. I've got very elderly patients, most of them with endocrine disorders, whose old age I help to extend with a better quality of life. We've got dogs of seventeen or eighteen, and some cats of up to twenty. Older than Methuselah, and yet here they still are, thanks to our care. We learn from them."

"What do you learn?"

"Our minds very often don't help us to accept the limitations inherent in ageing, which stops us seeing that, albeit with some ailments, there's life in us still. Animals adapt to these limitations, they don't drive them mad like they do us."

Arsuaga pipes up, having heard part of our conversation: "Social animals, in nature, have a hard time of it when they get old. The young ones give them a hard time. When lions, wolves, or bison start to become hunched and wizened, the young ones give them hell. Hence why they don't live very long."

"We vets," says Lola, "help animals to age with dignity, but we don't apply the same philosophy to ourselves. There's no reason to feel bad about being eighty years old and on seven medications."

"My dignity does not reside in my prostate," says Arsuaga. "Though that would be good! Nor in my eye lens or my liver. I don't know where it resides, but not in my prostate, that's for sure."

There's a silence that I — as the oldest of the three of us — find uncomfortable and which the palaeontologist concludes in a jovial tone: "Right, that's enough decrepitude and death for one day. Let's go and have a look at love."

Love is not very far away. It turns out to be in a facility for horses, where Arsuaga introduces me to Mónica Domínguez, a mare gynaecologist and horse urologist, and to Paloma Forés, vice-dean of students.

"I've told you before not to introduce me to so many people," I tell the palaeontologist, having pulled him aside. "I get muddled with so many names. I get muddled with so many characters."

"Life is full of characters," he replies.

We're in a large room, very large, in fact, with high ceilings, a kind of indoor racetrack, where a huge stallion (or at least he seems huge to me) by the name of Nervudo soon appears, led by an expert student.

"This horse," says Mónica Domínguez, "is an eternal teenager. He's seven."

"And how long does a horse live?" I ask.

"Up to thirty," replies Arsuaga.

"Nervudo," Mónica Domínguez continues, "is an army stud horse, a Spanish thoroughbred. For size and proportion, he'd score a perfect ten in competition. I call him an eternal teenager because he's cocky. The technical name for a cocky horse is 'foalish'."

In an adjacent room, also quite large, Nervudo is awaited by a mare on heat, named Mexicana, who's being held in a kind of cage where her movements are very restricted. Next to this real mare is a fake one, built around what looks like a vaulting horse in a gym, which has been covered in something faintly resembling the body of a cow. I quickly sense that this is a crude sex toy, and they soon confirm as much: the idea is that Nervudo, excited by the presence of the real mare, will ejaculate on the fake one, so they can collect his precious semen.

Jesus.

And here he is now, whinnying with excitement and pleasure at the scent of the mare on heat; he knows he's here to fuck. With us in the room there's a group of students, about a dozen of them, all on tenterhooks for today's lesson.

After letting the horse bring his nose close to the backside of the mare, who has raised her tail receptively, the student leading Nervudo takes him to the fake mare, called a dummy or phantom (phantom!). Nervudo's erection, while considerable, still isn't, I'm told, at its full potential. When it does reach that limit, another student takes the gigantic member in her right hand to wipe it with a cloth that has been dampened with some antiseptic liquid she's holding in her left.

"That," Paloma tells me, "is to make sure it's totally free from any bacteria that could get mixed up with the semen."

Once the washing has taken place, another student appears

with an artificial vagina, an elongated contraption made up of a leather tube with a rubber receptacle inside containing a water circuit which is the same temperature as the mare's vagina. Nervudo then mounts the fake mare (the phantom!), and the student — not without some trouble — inserts the penis into the artificial vagina. The horse shifts around uncomfortably, stamping its hoofs.

"There's something he's not happy about," says Paloma. "Either the water temperature or the pressure. He's going to dismount."

The horse does indeed dismount, and his erection fades. His caretaker tries to calm him, stroking his neck while the student checks the vagina and goes to change its water, which wasn't at the correct temperature.

"It's got to be pretty warm," Paloma tells me. "Forty, forty-five degrees. Remember, thirty-seven degrees is like a lukewarm baby's bottle, but a vagina needs to be hotter."

The association of a baby's bottle and a vagina unsettles me for a moment, but we're here to learn something about love, or about sex, or masturbation, I'm not really sure, we're here to learn something we didn't already know, and I concentrate all my senses on that.

The horse starts to fidget uneasily; he must have smelled the real mare again, because his erection is even more flagrant, if that's possible, than before.

"All this," the vice-dean explains, "is a part of the breaking process. You've got to know how to handle a horse whose semen you're about to take; he has to obey you, even when he's got an erection like the one you can see here. A lot of problems, when you're trying to breed them, come down to a horse not being properly broken. This girl" — she's referring to the student who's leading the stud horse — "has been working with us for three years, she's coming to the end of her course. As you can see, she's got him under control, she's very good."

"She's got him wrapped around her little finger," says Arsuaga.

At that moment, the student comes back with the artificial vagina. The horse mounts the phantom again, and they insert his penis into the artificial vagina. Nervudo stamps, stirs, shoves, bites the dummy's artificial mane …

"It shouldn't be taking this long," says Paloma, some concern in her voice.

But the horse is looking desperate because, they tell me, he hasn't got the vagina in a comfortable position.

"It's too much to one side," adds Paloma.

"He's not having a great time," I suggest.

"Yeah."

"Do they take long to ejaculate?"

"A minute, a minute and a half," she says.

"Doesn't seem very long," I say, trying not to sound conceited. "Is he a premature ejaculator?"

"No, it's not premature. You've got to remember the whole part before the mounting, the foreplay, the courtship. Rabbits are more premature than that."

"And when a horse mounts a real mare, is there time for her to enjoy it?"

"Mares … well, there's no sign that they have orgasms as such, but what we do know is that when they aren't on heat, they reject the male entirely. They don't climax, if I can put it like that, but they do *like* to be mounted."

"The problem with artificial insemination," I add, feeling sorry for Mexicana, who is looking desirous, "is that now the poor females don't get to enjoy any of it. A hand puts the semen inside them, and they don't even know."

"The inseminator's arm is just like the penis," Paloma points out. "It reaches as far as the cervix too, and after inseminating her, it too stimulates her, because the male, when he thrusts, pushes on the cervical canal … But no, it's not the same. When they come into contact with the males, oxytocins get released and then they're more fertile."

After a few sterile thrusts, Nervudo gives up again, to the distress of the bystanders.

Mexicana, in the meantime, trapped inside her steel structure, lifts her tail while opening and closing her moist genitals, as if blinking.

"That," says Paloma, "is what we call a wink. She's winking. It means she's very receptive."

"Poor thing!" I exclaim, genuinely moved.

At the fourth or fifth attempt, after a professor has taken charge of the real phallus and the fake vagina, after a minute and a half, more or less, of back and forth, Nervudo manages to ejaculate.

There is a general sigh of relief around the room. Personally, I'm not sure I've witnessed the lovemaking I was promised. I almost prefer Thanatos. The whole thing was somewhere between the satirical films of L.G. Berlanga and plain cruelty. I really felt for Nervudo, but also for Mexicana. I'm drained, as if the amatory effort had all been on my part, and it was also nearly lunchtime, and those are two circumstances that always cloud my thinking.

The professor who took over approaches us with the transparent plastic receptacle containing the horse's sperm. I don't think it's all that much for such a big beast, either.

"How much do they ejaculate?" I ask, trying not to sound competitive.

"Between forty millilitres and a little over a hundred, depends on the horse," one of them tells me.

Then they put a drop of the precious liquid on the slide of a microscope connected to a monitor so that the students, teachers, and those of us sitting in can look in amazement at the dance of the thousands or hundreds of thousands of spermatozoa whizzing about in search of somewhere to go. There are some that move upwards and others that spin round in circles, as if confused.

"Those ones going round and round," says somebody, "are no good."

"Maybe they have an obsessive personality?" I venture, to no response.

If I were a horse sperm, I think to myself, I'd belong to that self-reflexive, more-or-less-useless category, too.

Back home, the palaeontologist reminds me of the subject of cell proliferation.

"You can explain it to me some other day," I plead with him. "Now's not the time."

FOUR

Let us be Epicureans

I was going through one of those moments in your existence where you feel you need to do something to survive. To survive the days, the hours, the moments of getting out of bed and getting into it, which are the most dangerous of the day, like take-off and landing for a plane, or startup and shutdown for a computer. I didn't know what "to do something" meant, since the ideas that ran indistinctly through my head were to sign up for a yoga class, to go back to the exercise bike, to reconvert to Catholicism or join a Satanic sect.

Something.

Do something.

Finally, I thought I'd lose weight.

Not that I was fat, or not very fat. I weighed eighty-three and a half kilos, which, for my height (1.75 metres), was only slightly overweight. But I started thinking my unease was the result of those few extra kilos, the way people start projecting the hatred they feel for themselves onto some public figure. Given that this excess was a part of me, hating it was a way of hating myself, a subject in which I am an expert. I would look at myself in the mirror and carry out imaginary trims to this or that part of my body.

What would I do with the extra flesh?

I would bury it in the garden. Everyone's got a skeleton in the garden, or in the closet, but it's more hygienic to keep it in the garden. I imagined the police showing up at my door with a warrant: "We're going to need to search your garden."

And they'd spend two or three days making holes here and there, though especially in places where the lawn had grown the most, until they came across my seven or eight surplus kilos, in a state of decomposition by now but perfectly adequate for taking DNA samples that would inevitably match mine. The normal weight of a newborn baby is three kilos. Which means I would have disposed of two babies. I had been carrying within my body the weight of two newborns, maybe one of them corresponding to the twin I've always believed I consumed in my mother's womb and for whom I still hadn't been held accountable. Because that's where I come from, from cannibalism, like any self-respecting Neanderthal.

But anyway.

Where to begin?

I was told about a very famous dietitian who was often on the radio and whose results everybody responded to with full-throated praise (I hate that word, "full-throated", because even as I say it I can feel my own throat swelling, making it hard for me to breathe).

Everybody was responding with full-throated praise to the work of that dietitian, whom they also referred to as a nutritionist.

I managed to get hold of the number for her office, and one Monday in early March when I found myself on the verge of suicide, I called.

"Doctor X's office?" said the voice at the other end. "How can I help you?"

"I'd like to schedule an appointment to see the doctor," I said.

"And what is the purpose of the appointment? Is it for slimming or for dealing with some digestive problem?"

For a moment, I just sat there feeling flabbergasted.

"To be honest," I said, "it's not for either of those. I'm not ill, and I'm also not sure I have a terrible weight problem. I'd like to talk to the doctor to study the possibility of introducing healthier nutritional habits into my life."

"Sorry, sir," replied the friendly lady at the other end of the line, "you'll have to choose one or the other: weight loss or being ill. If you choose weight loss, the first consultation is a hundred and thirty-five euros and lasts an hour and a half."

I preferred not to find out how much the being-ill option would cost.

"I'll have to think about it. I'll call back," I concluded.

After I had recounted this odd capitalist experience of mine on the radio, a friend called to give me the number of a nutritionist who lived in Barcelona and who offered sessions on Skype.

I called her right away, as I was longing to introduce something new into my life. I remember what I said: "There's one thing that's non-negotiable, and that's wine. I drink half a litre a day with my main meal. And one or two gin-and-tonics, a couple of evenings a week."

"How about we negotiate?" said the friendly dietitian. "Would you agree to drink only red wine, a decent vintage, or failing that, champagne?"

"I would," I said.

"And could you drink a quarter of a litre instead of half a litre with your meals?"

"I'll try," I replied.

Not long after starting this diet, I received an email from Arsuaga, who seemed to be watching me from a distance, like my mother, from whom I was never able to keep anything. He was inviting me to join him for a gargantuan dinner, all seafood — which is what I like most — during which he would be explaining something that he said he'd rather save until then.

I hid the fact that I'd just gone on a diet, and we agreed to meet on 25 March.

"Before dinner, though," Arsuaga added, "let's take a walk in the Fuente del Berro park."

"Why?" I asked.

"To work up an appetite."

I didn't need to work up an appetite; on the contrary, I've already mentioned I was on a diet, and I was already hungry every hour of the day for the things I'd been deprived of, but I pretended it all sounded just fine.

In the Fuente del Berro park, there are peacocks that court in the spring. The palaeontologist and I stop in front of a male and a female — peacock and peahen — who are standing some ten metres apart. The peahen is staring into the distance and the peacock is watching her staring into the distance, just like when you see a painting inside a painting. The peacock's stare is a meta-stare.

"Sooner or later," says Arsuaga, "the peahen will look in the peacock's direction and something will happen."

Soon, the peahen does indeed turn her neck, and notices

or pretends to notice the presence of the peacock, who unfurls his tail into a fan as if somebody has flipped a switch. After a few moments, the peahen goes back to her previous position, and the peacock puts away his tail again. The peahen's action is repeated several times, prompting successive automatic movements of erection and deflation in the peacock's tail, like when a child plays at turning a light on and off. Finally, the peahen takes her leave, and the peacock is left just standing there, perhaps trying to process his amatory frustration.

"It's not going to happen," I say to the palaeontologist.

"We don't know whether it is or not," he replies. "We should maybe have come earlier in the day. It's very hot now."

"It's never too hot for a fuck," I venture.

"You've really got no clue," he says.

We walk on through the park in search of some real mating, though that never comes to pass, even though we do witness quite a bit of the glancing and tail-raising game, not unimpressive in itself.

"These venereal failures," I say, "remind me of the frustrations of Sunday afternoons in my adolescence."

The palaeontologist ignores this.

"Darwin knew that for something to be truly scientific, laws had to be discovered and formulated. That's what Newton had done with universal gravitation."

"And did he succeed?"

"Of course. He discovered and formulated, firstly, the law of natural selection, and then that of sexual selection. We've already talked about this, but not enough. Bear in mind that biology, until Darwin came along, didn't have laws."

"Right."

"Peacocks are the perfect way to demonstrate the existence

of both types of selection. There are things about their bodies that have come about through adaptation to the environment, and others that are only there to make them sexually attractive. What are eyes for?"

"For seeing," I say.

"And legs?"

"For walking."

"And so on successively. There you have a series of physical features that come about through adaptation to the environment. But then, what's the peacock's tail for?"

"For screwing," I reply.

"i.e., for sexual selection. That whole crest of feathers on the male peacock's tail has no ecological function. Everything you see on this creature, from its beak to the base of the tail, is about the struggle to survive. The tail feathers are solely about reproduction."

"And in the female?"

"In the female, everything's adaptive. She's pure ecology."

"What luck!" I exclaim, while imagining how inconvenient it must be to drag a tail around, and casting a secret glance at my watch, as it's nearly time to eat and I'm starting to notice the drop in my both my sugar and hydrate levels, or whatever it is that drops in my blood at around this time.

"Within the struggle to reproduce," Arsuaga continues, "there are two models."

"And those would be … ?" I say, to speed things up.

"Battlefield and catwalk. When it comes to mammals, the fight is a literal fight. Whereas with birds, they fight on the catwalk. In both cases, the females mate with the ones who have the best genes."

Arsuaga stops, fixes me with an appraising stare, and asks

whether I understand the difference between battlefield and catwalk. I give an affirmative nod because, as a mammal, I have known the battlefield, and we keep on walking until we come face to face with another peacock who is filled with anxiety, because somewhere close by there's a female who won't deign so much as to glance at him.

"Poor thing," says the palaeontologist, "he's up to his eyeballs in hormones. The spring sunlight we're enjoying today increases hormonal secretion. Keep in mind that the offspring need to be born at the moment when nature's at its most plentiful. For birds, that means the time when there are most insects around."

"Most insects," I repeat, mechanically.

"Precisely," adds Arsuaga, "it's the one hundred and fiftieth anniversary of Darwin's *The Descent of Man, and Selection in Relation to Sex*. There's a strong link between sex and death — I imagine you don't need me to explain that."

"Oh, no," I say, pretending the relationship is obvious, since my water and sugar levels are no longer dropping, they're plummeting like a suicide off a cliff.

"Go on."

"Well," I say, "the French call an orgasm a 'little death' because the orgasm is what precedes the big death that happens, sooner or later, to the person conceived through this small fainting fit."

The palaeontologist hesitates, as if faced with a pupil who's trying to con him.

"We can go and eat, then," he concludes, at last, to my intestinal delight.

———

The restaurant is located on Menéndez Pelayo, across from the Retiro park. It's called Zoko Retiro, and it appears to be Basque, as Arsuaga is. He has friends, Basque or otherwise, absolutely everywhere. He tells me they've prepared a special menu for us, which he's designed himself; it's bound to be incompatible with my diet, I think.

To hell with the diet, I say to myself.

While they set our places and bring us something to drink, the palaeontologist explains to me that he's a neo-Darwinian.

"And there are three things," he goes on, "that neo-Darwinism struggles to account for. It's given us an explanation for adaptations, ecologies, and reproductive systems, but there are three things it still hasn't got to the bottom of: why sex exists, why death exists, and why altruism exists."

"And why they're not bringing our food …"

"Neo-Darwinism started looking at these things," Arsuaga continues, indifferent to my faintness, "in the 1940s, and it's still working on them."

"And why does sex exist?" I ask.

"Half your chromosomes come from your father, and the other half from your mother; this means that, when an individual reproduces, in passing on its genes, it renounces half of them. Your children only carry half of yours, and your grandchildren, a quarter."

"So sex makes you gradually disappear as the generations go on."

"Exactly. So why not go with parthenogenesis, say, the process by which the ovule splits, without needing to be fertilised — producing a clone? That's what happens with certain snakes and lizards. Why are we not parthenogenetic? What explanation is there for sex?"

"We could ask the same of parthenogenesis, as well as sex …"

"The thing is, sex, looked at in simple terms, has no advantages. Why would you want to give up half your genes? Think about salmon."

"If I start thinking about salmon now, I'll expire from hunger."

"Better still, do think about them — it'll get your gastric juices going. So, the female salmon spawns, producing a million eggs, or however many it is. Why don't they just turn into salmon fry without the need for the male to come along and fertilise them?"

"That's the question I've asked myself so many times: who are we fucking *for*?"

"Why, biologically speaking, do the females need us?"

"If you know, why don't you just tell me?" I say.

"A romantic like you, a Kropotkin, a new-age kind of person, a hippy — you'd say there's an order to everything, that individuals are in service to the general objectives of the world. But I have to disappoint you: there is no answer."

"Oh, come on!"

"If there were, it would go hand-in-hand with the answer about death. Why do some species die at a certain age, and others at a different age? Who's writing the program, and why?"

"For the good of the species."

"In the Darwinian view, there's no such thing as the good of the species. In Darwinism, only the interests of the individual count. Natural selection follows the Scottish economist Adam Smith: the invisible hand of the market."

"Well, it's really the other way around," I point out. "It's Smith who plagiarises natural selection."

"As you wish," the palaeontologist continues, "though

Smith predates Darwin. The point is, the idea of the good of the species has lodged itself in our brains because we've heard it so many times. All the dramas we see unfolding in nature have been explained, always, with reference to the good of the species. I call this the 'Félix Rodríguez de la Fuente discourse', because it limits itself to the kind of reasoning that in its day seemed beyond question. But what the hell does evolution care about the good of the species? It might matter to nice, well-intentioned people like you, sure, but evolution couldn't care less. In other words, as things stand, there's the same explanation for death and for sex: no idea. So don't let it get to you."

"I'm not, but they could give us a little something to nibble on."

"Darwin," Arsuaga continues, unfazed, "has been opposed by every one of history's progressive new trends. He was criticised for belonging to Victorian society, in which, the supposition is, everyone looked favourably on market competition, just as they did industrialisation and colonialism. A whole set of theories can be provided for why Darwin was a total son of a bitch. Except he wasn't. He was anti-slavery, he was against colonialism and vivisection. He lived very simply in a small town. He was anything but a Victorian capitalist. I'm not going to say he wouldn't have unconsciously assumed the values of his time, or that these didn't have some bearing on his scientific theories ..."

"Naturally."

"... but I think scientific theories should be judged for their scientific value. Ideologies shouldn't come into it. The periodic table is the same in North Korea as it is in South Korea, it's universal, and science has to aspire to universality. Sodium has the same positively charged ion in communist China as it does in the United States."

"Something similar happens with alphabetical order," I say. "All orders change except alphabetical order. It never even occurred to Franco to decree that the alphabet should begin with an F."

"Well, there you are. Our problem is our search for beauty; we have this need to find some beauty in death. Dying in order to make way for others, for the good of the species, or Pachamama, or whatever other new-age nonsense, gives us a feeling of tranquillity. But I'm going to tell you, as clearly as I can: natural selection is death, it's death killing children. Even more than that: natural selection is infant mortality."

"Do you know what Rilke said about beauty?" I jump in to avert the impending monologue.

"What?"

"That it's just the beginning of a terror which we are barely able to endure."

"It's well put."

I don't think the palaeontologist is aware of the desperation he exudes when he refers to death.

"But, anyway," I add, "from what you're saying, and if it's right that death doesn't exist for the benefit of the species, we can deduce that it exists to benefit the individual. Weird, right?"

"Well, there's the challenge. The simple conclusion is that you die so your children can inherit your apartment. We can understand that, because it's intuitive. But science, I've told you a thousand times, is *counter*intuitive. The part that's difficult to explain is that death benefits the dead person."

"Does it?"

"I am fundamentally someone who provides questions,

not answers. But we'll see. We'll see about this, and about sex, altruism, and cooperation, too."

At this moment, the waiter appears and puts a dish of shelled peanuts on the table, and beside it he places a bowl of pesto. He also brings us some chopsticks because Arsuaga says this appetiser you eat with chopsticks.

In my diet, dried fruits and nuts are completely forbidden, because of their calorie count. Which means that Arsuaga seems to know everything that isn't good for me. My religious sense (from *religāre*, to join) isn't suited to the total meaninglessness of existence that neo-Darwinism preaches, and my stomach isn't suited to this oily legume. But I'm starving, so I take the chopsticks, pick up a peanut, dip it in the sauce, and lift it to my lips. A divine mouthful.

"Doesn't the way chopsticks move remind you of anything?" asks Arsuaga.

"What's it supposed to remind me of?"

"The beak of a bird when it's eating."

He's right. I'm good with chopsticks, due to the fact I adore Japanese food, so I click them together, imitating the movements of a canary's beak.

"When I designed the menu they're going to serve us," Arsuaga explains, "I asked for this to be the starter."

"Has it got anything to do with death?"

"No, but it is funny: José Antonio Valverde, who, among other things, designed the Doñana reservoir, was a naturalist with a broad perspective on things, and a great draughtsman. As someone who didn't specialise in evolution, his vision of it was far broader. Because it wasn't his job, he came up with things that never would have occurred to a professional."

"What things?"

"For him, grain is at the origin of human evolution. You get lots of people who think human evolution can be boiled down to a single variable."

"That's the dream of religion," I say. "Religions are systems that explain everything."

"According to the granivore hypothesis," Arsuaga goes on, "all species that feed on grains, on seeds, have split masticating apparatuses that are separated by a space, the anatomical name for which is 'diastema'."

"Masticating apparatuses," I repeat, because it sounds nice.

"They're called this," Arsuaga explains, "because science is awash with the mechanistic conceptions from the seventeenth century, which was when it came into being. To Descartes, everything's a machine."

"Hence also the musculoskeletal system, the circulatory system, the digestive system, the genito-urinary system, et cetera. We're made up of mechanical systems."

"Look at a mouse: it has the incisors at the front. Then there's this big gap."

"The diastema," I repeat, bringing to my mouth, with the tips of my chopsticks, a single peanut dipped in pesto that I allow to sit, before chewing, in my own buccal cavity, on my tongue, to give pleasure to my tastebuds.

"Let's say that a mouse's incisors serve as pincers, and the molars as grinders. Write this down," says the palaeontologist, pointing his chopsticks at me. "Pincering and grinding. Birds don't have teeth — their predecessors did, but they've lost them along the way. Which isn't to say they don't have a grinding apparatus. In them, it's the gizzard, which is this very strong muscle, very tough, which can reduce a grain to powder. An insectivore has no need of such grinding, it feeds itself on insects. A carnivore has no need of such grinding, it just tears the meat off and swallows it in chunks. But granivores, they've got to be able to pulverise things. Partridges have developed a muscular structure in the stomach, the gizzard, which is the real grinding apparatus. They sometimes swallow little stones, to help with this grinding process. Dinosaur skeletons have been found with stones polished by the gastric juices in their stomachs. They're called 'gastroliths'."

"Gastroliths, that's a great name, you can just see its etymological body right there in full view," I say, as I bring to my mouth another peanut, which I'll use to test the skills of my masticating apparatus, while simultaneously becoming aware that I possess a diastema in which I could, if I so chose, store more than one grain.

"In poultry and gamebirds, the pincer is easily distinguished from the grinder," says Arsuaga.

To which I reply, enlivened by the calories supplied by the peanuts and the pesto: "Chicken gizzards, with rice and vegetables — they're delicious. I remember having had them in Colombia and Ecuador."

"So then," continues the palaeontologist, while continuing to use his pincer and his masticating apparatus, "what happened to humans when they climbed down out of the trees?"

"What?"

"They became granivorous. Their hands were now no longer needed for locomotion, and they started using their fingers as pincers to pick up grains. That's the moment when pincer and grinder became separate."

"The pincer would be the hand, and the grinder, the teeth?"

"Exactly. Our mouths become less pronounced, we stop having a snout. As for the incisors, they're very small.

What we've really acquired here are grinders. We've become hominids. When a chimpanzee eats fruit that's ripe, it uses its incisors to cut it up and then swallows it in chunks, barely chewing at all, because when fruit's ripe, it's soft and doesn't need to be ground up. This is why chimpanzee molars are small in comparison with the ones we see in our Australopithecus ancestors, who were the first bipeds. What I've just summarised for you here — the idea that you'll find grains at the origin of human evolution — is Valverde's granivore hypothesis. He spent his life convinced he was going to win a Nobel Prize."

As if we were within a theatrical performance directed by Arsuaga, no sooner is the granivore hypothesis concluded (along with the pesto peanuts) than the waiter appears — his name is Luis Gallo and he wants to be in our book — and gives each of us a plate with a roasted tentacle of octopus accompanied by an excellent vegetables garnish.

"The octopus," says Arsuaga, brandishing his knife and fork, "means we can now address the question of pre-programmed or un-programmed death."

"So let's address it," I agree, cheered by the culinary delight that has appeared before me.

"We've already commented that different species have different lifespans, as though this was something that had been pre-programmed. An octopus lives to two or three. First of all, it starts to lose that marvellous skin it's got, then it falls apart."

"So we do have a case of decrepitude there, then," I point out. "But you said that in nature you're either in your prime, or you're dead."

"But it's so brief, it can't be called 'old age'. It borders on suicide. They die in an instant, as soon as the job of protecting the fertilised eggs in their lair is done. That takes a month, sometimes a few months, until the eggs hatch and the offspring all disperse. Do you know the biological term for when species die as soon as they've reproduced?"

"What?"

"Semelparity. The reason being, Zeus had a mortal lover named Semele, and Hera, his wife, was very jealous. Hera tried repeatedly to convince her husband's lover to leave him. But Semele, who was pregnant by the god, wouldn't listen. Now, when Zeus had relations with Semele, although he did not hide his godliness, he took the form of a mortal. Hera therefore insinuated to Semele that she was being tricked, that her lover wasn't a god at all. 'The next time you see him,' she added, 'ask him to show you his true nature, then you'll see I'm not lying.' Semele did so, asking Zeus to appear before her in all his splendour, but he was resistant, refusing over and over again, until, worn down by her insistence, he allowed her a glimpse of the thunderbolts and fire of which he was composed. Semele was instantly reduced to ash, but Zeus managed to rescue the fetus she'd been carrying, and implanted it in one of his thighs, to allow it to reach term. Some months later, Dionysus was born, god of the grape harvest, god of wine."

"You amaze me," I say, before bringing to my lips — in tribute to Dionysus — the glass of verdejo I've just been served.

"So that's the story behind the term 'semelparous' — animals that die as soon as they've reproduced."

"That's great!" I exclaim. "That thing of implanting the fetus into the thigh, as if it was a graft! And such astonishing intuition when it comes to advances in contemporary genetics!"

"Semelparous animals," Arsuaga concludes, "forego the responsibility of raising and educating their offspring."

"They don't even get to meet them."

"Practically all fish, amphibians, and reptiles abandon their fertilised eggs and forget all about their offspring, they don't look after them. Sea turtles just leave their eggs buried in the sand and get back in the water, but not birds, birds do the incubating, they feed the chicks. But this isn't the point. The point is that semelparous animals die the very moment they've reproduced."

With the octopus course now dispatched, Luis Gallo appears with a couple of small delicacies: two portions of salmon tartare, each on a slice of lime. They are to be eaten in a single bite, sinking your incisors slightly into the lime. The citrus cleans the palate, leaving it clean to pick up the taste of the fish.

"I asked for it to be *salmon* tartare, because salmons are also semelparous," says Arsuaga.

"That's true," I remember, "they die when they spawn."

"Both sexes. Both die."

"But only one of them spawns."

"The female provides the eggs, the male provides the sperm; they do it simultaneously, then they both die."

"So well synchronised! And they just leave their offspring there, with no one to look after them."

"The female makes a hollow for the eggs. Have you never seen anything on TV or in the cinema where they show scenes of Alaskan rivers, with dead salmon scattered everywhere? They all die, not a single one makes it back to the sea. The ones you see in the documentaries, those are Pacific salmon. Our salmon, the Atlantic kind, can make it back to the sea, with a little luck, so they get to have another go at swimming upriver the following year. But you can't get Pacific salmon in a Spanish restaurant."

"The salmon you get in documentaries are programmed to die the moment they pass on the baton. They do the business, then perish."

"There's your inner Kropotkin again!" protests the palaeontologist. "Next you'll be saying that nature knows what it's doing."

"It's a mechanical, intuitive response," I say apologetically, though by now I should have learned that science is counterintuitive. "But I'm not a scientist, sorry."

"Don't worry. None of us is perfect: I'm just a scumbag Epicurean. Pass the wine."

"You're dis-Kropotkinating me. Thanks to you, I accept that it's the earth that goes around the sun, even if my perception tells me the opposite."

"Like it says in Quevedo's love poem — the greatest one ever: 'They shall be ash, but they will have meaning; / dust they shall be, but dust that's in love.' There's got to be some meaning to death, right? Can you even live without trying to find a meaning for everything?"

"Perhaps," I concede. "I do think humans are fundamentally seekers of meaning. But we've talked about that before. Answer me this: why do salmon atoms behave like that, and ours reproduce more than once, as often as we have children?"

"For the good of the species," replies the palaeontologist with an ironic smile.

"The salmon doesn't need to die for the good of the species."

"Exactly. Once it's carried out the fertilisation, producing the next generation, its mission on mother earth is complete. It gives up its life, happily, with great pleasure, having completed its task in the universe. That's what your 'meaning seekers' say, anyway. Whereas for an Epicurean like me, there's nothing,

nothing has any meaning, only the search for pleasure. And I hope you know the proper meaning of the word 'pleasure'."

"You sound so crestfallen when you say it that I do understand, and though I don't know why, I'm reminded of that Unamuno title you're always quoting: *Tragic Sense of Life*."

"But was that salmon tartare good or what?" he says, steering us away from tragedy.

"Excellent, though I don't know why it bothers you that we think about the meaning of life."

"It's just that, for an Epicurean, there's nothing more than molecules, atoms; Epicureanism is pure science. That in turn means the search for a state of mind that's serene and open to life's pleasures."

"Well, if you ask me, there are days when your state of mind doesn't seem very serene, or very open to the pleasures of life."

"I'm working on it. I'm trying to master ataraxy, which is a form of equilibrium between the mind and the senses."

"A kind of unflappability in the face of adversity," I say.

"Put it however you want."

"I'm up for that, too, ataraxy."

"But you think things happen for a reason," Arsuaga insists, determined to eject me from his club, "and that's incompatible with Epicureanism."

"I'm always going to disappoint you, because I'm nothing more than a poor Neanderthal. And I don't deny that it's all just atoms randomly crashing together. I'm just saying that belief isn't incompatible with trying to explain why there are species that die at the exact moment they reproduce for the first and last time, and others that survive to reproduce again in the following season."

"The only convincing explanations are the mechanistic ones."

"But it's not wrong to wonder about the laws of nature."

"You're very welcome to. The scientific tradition consists of doing precisely that."

"If a student asked you why there are species whose individuals disappear the moment they give way to the next generation, while others don't, what would you answer?"

"Well, I would try to give an answer, but the real answer, to begin with, is 'I don't know'."

"And beyond that 'I don't know'?"

"Beyond that, we're into speculation."

"Go on, then."

"Sex, death, and altruism. Those are the three problems faced by neo-Darwinism."

We're interrupted by Luis Gallo, now bringing us some more delicacies, each served on bamboo leaves. They are tuna bonbons, filleted and flambéed, accompanied by a range of sauces, some of them very spicy, along with wasabi pearls and some seaweed that, I will soon confirm, tastes of barnacles.

"Sophistication and efficiency," says Arsuaga.

"I agree," I say.

"Let's get another bottle of verdejo."

I think about my diet, which has gone totally out the window, and this gives me a feeling of unease that disappears as soon as I attack the tuna (filleted and flambéed). For a moment, thanks to the sweet feebleness prompted by the alcoholic and alimentary wellbeing, I'm about to confess to the palaeontologist that I've gone on a diet. But I hold back because I'm not sure it would be a very Epicurean decision.

"We were saying," says Arsuaga, "that salmon reproduce and then die because they're semelparous. Not you, you're iteroparous, which comes from the Latin *itero*, to repeat, and *pario*, to beget. In other words, you're able to reproduce a number of times before you die."

"Which causes many problems of an economic and emotional nature. But what would life be without problems?" I say, before tasting the wasabi pearls.

"Enough of the sentimentalism. Do you know why I asked them to serve us this dish on bamboo leaves?"

"Not yet, but I'm starting to see that this food, unlike life, is filled with meaning."

"Bamboo," he explains, "doesn't generally reproduce through seeds. It's a strong grower, very invasive, but it grows by the roots, resprouting over and over again. You could even take a bamboo stem and move it to a different continent, and it would start to grow again, spreading far and wide, if you didn't control it. It isn't a tree, it's a grass — graminaceous — even though it's woody."

"A tough, prehistoric sort of grass."

"It lives for a hundred or a hundred and fifty years; then one day, suddenly, it produces flowers and dies. There we have a problem of sex: why bother to flower and interbreed, given how well it was getting on just cloning itself? Why, suddenly, does it need a kind of pollination that it's done without for decades?"

"Maybe to avoid inbreeding?" I suggest.

"Inbreeding isn't something that registers with biology," Arsuaga points out.

"But we know it's destructive."

"We do, but nature doesn't have eyes for the future. It doesn't look ahead, it doesn't plan. Write this phrase down, it's one of mine: 'Evolution doesn't seek, but it does find.'"

"Done."

"Evolution has no particular aim in mind, but ... Let's be

clear: does bamboo commit suicide? No! That's a complete mistake. It flowers then dies, like things that are semelparous, though that term is only used for animals."

"Pre-programmed death?"

"Do you know what an agave plant is?"

"Yes, it's the one tequila comes from, like a cactus."

"Right. It lives for a long time; then one day, out of the blue, a stem appears with flowers on it: the plant bears fruit, then it dies."

"Like bamboo."

"Bamboo likely produces flowers when circumstances are favourable. Maybe it needs to wait one hundred years for that. But when it does, it reproduces, and then dies. Why?"

"Because nature is wise?"

The palaeontologist laughs and brings his glass to his mouth, adding sarcastically, "Right, Pachamama and all that — Mother Nature."

"What is it, then?"

"Natural selection, the reaper scything all and sundry."

"If there's no one conducting the orchestra, why does life not collapse?"

"An Epicurean would say it's because it's a perfect machine that's built itself."

"A perfect machine with some species that die on giving birth, others that give their children a good start in life, some that live a month and others seventy years … And all this chaos, which functions like clockwork, has been constructed by natural selection."

"That's it, and then some. That's natural selection."

"I'm inclined to accept that, Arsuaga, but I'm having a lot of doubts."

"Stop trying to find the meaning for it all. Become an Epicurean, enjoy what you are. You've been given this gift of consciousness, which is one of the rarest products in all of evolution. There isn't anything else. Don't let yourself be manipulated, and that's an end to it."

"This business of having been 'given a gift' all sounds a bit Félix Rodríguez de la Fuente to me."

"I'll accept that. We are physical matter, but the lucky kind. Every other creature lives, but isn't aware of it. You're capable of living with your eyes open. There is no meaning, no Pachamama, there's nothing else."

"So let's come up with a narrative about the lack of meaning."

"Not now, I can't."

"But there *ought* to be a narrative about the lack of meaning."

"I'm offering you the chance to save yourself. To stop seeing cooperation where in fact all is violence; to stop perceiving some chivalrous combat where it's all actually just a fight for survival; where, though aggression is the only thing going, you ask for forgiveness; where, though death is all there is, you somehow see self-abnegation and sacrifice."

"So where I see violence, what do you see?"

"You're not capable of seeing violence, you're a kind of hippy. I'm the one who sees it."

"But isn't violence really just a projection? The fact that a cheetah eats a gazelle isn't violence, unless you project your own human feelings onto the act."

"Our readers will enjoy it greatly if you manage to convey a pleasant version of biology."

"Our readers will enjoy it greatly if we challenge their preconceptions," I reply.

"I say it again, I'm giving you the chance to save yourself. Write this down in your notebook: Arsuaga gave Millás the chance to save himself."

"You talk about salvation like the Christians."

"I've given you the chance to get out of the lynching, of being stoned to death. I've given you the chance to make *me* the one who gets stoned to death."

"You refer to the lack of meaning with a brazenness that's positively existentialist. Being an Epicurean does not, for example, prevent you from suffering over the fortunes of your nearest and dearest."

"Chance'd be a fine thing!"

"There are edges, then, emotional edges."

"It happens to me just like it did to Darwin. When his daughter died, he said, 'I'm done.'"

"What happened?"

"She was called Annie, she was ten. She was the apple of her father's eye. She fell ill, they think with tuberculosis. They took her to a specialist hospital, where they gave her the kinds of treatments that were used in those days, very cruel treatments that included ice-cold showers. When she died, Darwin said: 'That's it, I'm done.'"

"How did that change his thinking?"

"In every single way. There's no meaning …"

Luis Gallo is back, this time with a couple of small pieces of toasted brioche, on which they have put a small portion of caviar.

"Are these real sturgeon eggs?" I ask.

"Of course, what did you think?" grumbles Arsuaga. "When I come up with a menu, I do it properly. Sturgeon are very interesting because they can live to over a hundred."

"And how often do they reproduce in that time?"

"Yearly."

"Eating caviar makes me feel sorry, and guilty. It's so expensive ..."

"Don't hold back. Put it in your mouth whole."

I obey while shutting my eyes, to enjoy it as a solitary, intimate taste. The balance between the taste and the texture (very creamy, like butter) feels perfect. When I chew it, it emits aromatic oils. While I'm enjoying the taste, I hear, somewhere far away, the voice of Arsuaga: "It has great nutritional properties. It's full of protein, fats, minerals ... And there's lots of energy to be had from it, too: almost three thousand calories per hundred grams."

The quantity of calories, given that I haven't managed to forget my diet altogether, puts me on my guard, but it doesn't sour the experience. Could the palaeontologist have found out somehow that I'm on a diet?

At this point, with the pleasure of the caviar, albeit intense, lasting almost no time at all, Luis Gallo approaches again. He has an antique plate in each hand, those iron and porcelain ones. On each of the plates, there's a lobster. Oh God, I can't believe it! This menu is starting to look like a seaside holiday.

"These iron dishes," says Arsuaga, "last forever. They might get chipped, but they never break."

"And this menu is a work of art," I say.

"It's an Epicurean menu."

"A menu without meaning."

"A *well-organised* menu, with syntax."

While we attack the lobster, Arsuaga expounds, "What we've got here is an immortal species. The lobster can live to one hundred and fifty. You can tell how old they are from their length."

"And how many times do they reproduce?"

"The same as sturgeon: annually."

I take small mouthfuls because this dish, similarly, is just as much about savouring the texture as the taste — though they are inseparable, the palate enjoys trying to disentangle them, just as our sense of smell enjoys trying to separate the various aromas that make up a wine.

"An old lobster," Arsuaga continues, "is a lobster that's in great shape. It doesn't experience decrepitude."

"This excellent food," I say, "is absurd, it's meaningless."

"It's meaningless, but it's also delicious."

"I'm all in favour of meaninglessness."

"Write this down: octopuses live for two or three years, whereas this guy lives to one hundred and forty. And both cases seem like planned obsolescence."

"But how do we explain that? Who is it that programs a species' longevity?"

"Each and every species has one, that's a fact."

"But you can't help wondering *why* it should be this species and not that, whatever the answer."

"If it worries you that much, just sign up to the Pachamama theories. Earth as intelligent organism, all that."

"The fact of the matter is, under all that variety, all that chaos, there really is an impressive unity."

"There's the *perception* of a system," corrects the palaeontologist. "The more elements an ecosystem has, the more complex it is. Greater complexity, a greater number of properties. If we could teach politicians about systems biology, it'd be a fine thing, because the more complexity there is, the more potential. That's the premise for systems biology. Let's ask for dessert, though, maybe we'll find some answer there."

"I'm not sure I need your answers. This meal has been an answer, I'm not sure to what, but that's what it feels like: like an answer."

Luis Gallo approaches now with a couple of plates with miniature goat's cheese pancakes accompanied by some red berries that are aesthetically impeccable.

"The meaning of this dessert is in the red fruits," says Arsuaga. "They're antioxidants. Remember that old age is a product of oxidation."

"I take antioxidants," I confess.

"Congratulations, you've become a slave to the pharmaceutical industry. Honestly, Millás, you'll believe anything."

"That's very true. But what's wrong with taking antioxidants?"

"Consider the mouse, the poor mouse, which barely gets three years of life. So what's going on? Why, over three billion years of evolution, hasn't it been able to produce the same kind of antioxidants as a lobster? Are mice just total dunces, or what?"

At that moment, Luis Gallo drops a glass, which crashes onto the floor and smashes, very noisily, very close to us. Arsuaga turns to him, and says: "Glasses break."

"They do, yes," says Luis Gallo. "If it's not a customer dropping things, it's me."

"How long does it take," asks Arsuaga, "between buying a batch of glasses and half of them getting broken?"

"If I buy one hundred glasses today, six or seven months from now, fifty of them will be gone," replies the waiter.

"But not by getting worn out," the palaeontologist insists, "but because of breakages."

"Of course, because of breakages. None of them ever die

through wear and tear. They never last long enough for that."

"Fine, thank you. Can we have the coffee now?" Arsuaga asks.

Then, looking at me calculatingly, as if to ascertain whether I am still lucid despite all the alcohol and all the food, he adds: "And there's your answer. In the 1950s, Sir Peter B. Medawar, winner of the Nobel Prize in Physiology or Medicine, gave the neo-Darwinian explanation for death and old age."

"Which is?"

"He did it using test tubes, but I'll just use wineglasses. What was the lifespan of the test tubes in his laboratory — which, without exception, died because of breakages? Because neither test tubes nor wineglasses die of old age. What was the average lifespan?"

"It was … what it was," I say.

"But all the test tubes ended up dying in the same way as the wineglasses in this restaurant. The wineglass is immortal, it doesn't age, and yet it has an average lifespan, just like radioactive elements do. Imagine you're eternal. Would you still die?"

"I guess so, like Medawar's test tubes and the wineglasses."

"Exactly," says Arsuaga. "In nature, as we've said, there's no old age, no decrepitude, you're either in your prime or you're dead."

"Why?"

"Because in nature, things die in accidents, because of infections, parasites, starvation, or deprivation. They all fall sooner or later, like the test tubes and the wineglasses. There's no such thing as chronic illness in nature. Animals just don't grow old enough to develop such illnesses, which are the preserve of humans. Most chronic illnesses occur after people turn sixty, and are related to cardiovascular or respiratory issues, or degenerative processes in the nervous system. Such illnesses simply don't exist in nature, because nobody there gets old."

"Right," I say, trying to take it in.

"It's completely pointless asking how a wineglass will be doing in five years' time, because it's never going to grow that old. Every species has an average lifespan. Lobsters are strong and very few things eat them, thus their average lifespan is longer — it's a tougher kind of glass."

"A Duralex glass, let's say," I suggest.

"And what happens with octopuses?" continues Arsuaga, animatedly. "It's a kind of predator, a mollusc without a shell, because it lost that over the course of evolution. It spends its life down on the seabed, it doesn't even swim around. It's very vulnerable, very exposed — since it needs to move around to get things to eat. Octopuses are an extremely fragile kind of wine glass. You only have to look at them and they break."

"Okay, I'm with you. But why does it die when it reproduces?"

"We'll get onto that, I'm not sure when, but we will. Now, listen to this part carefully: if any animal in nature came down with one of the illnesses we call 'chronic' during its normal lifespan, it would die, full stop. Or it wouldn't survive, which is effectively the same thing. Suppose you have a gene that mutates, leading you to develop diabetes at the age of fifteen, during puberty. What happens then? What happens is you die before managing to reproduce, and nobody inherits this mutant gene, because it dies with you. In other words, natural selection has detected it, and eliminated it, by eliminating you before you've managed to reproduce. But now imagine that this mutant gene, which manifests as diabetes or some other illness, only shows up at the age of forty, once you're fully developed.

That means you'll have had time to sire offspring, although not to raise them all, which might mean you having fewer children than other people. In other words, natural selection will still eliminate it in the end."

"Let's go on."

"Let's go on. Now imagine that this mutant gene is expressed at the age of eighty, in a world where people generally don't live beyond seventy. The gene stays invisible to natural selection, it doesn't come up on its radar. And that's that."

"So what happens with humans, then, since we're a domestic animal?"

"Because of all the ministrations that comes with domesticity, humans go on living when all the wineglasses are broken, that is, when they should by rights be dead. Their predecessors have been accumulating all these genes, ones that are only expressed later on in life and are now being activated because they haven't been detected. This — the invisible genes becoming apparent to natural selection — is what we call 'old age', something that only we — and our pets — suffer from."

"You amaze me, Arsuaga. That is very convincing!"

"You, at seventy-five, shouldn't still be alive," he replies.

"Thank you."

"And since you shouldn't still be alive, you can't ask natural selection why it's failed to eliminate these genes that are tearing you apart, because all natural selection would say is, 'You, sir, should be dead.'"

"So what we call 'old age' is the accumulation of genes that have not been detected by natural selection, and it hasn't detected them because …"

"… practically all the individuals in the species died before that time."

"In other words, if we can surpass the average lifespan of our species by using medication and getting specialist care ..."

"... all these genes that natural selection never detected will end up having their say."

"I ought to be dead. And so my ailments are the product of an escape."

"Let's just pause for a moment with Medawar and his test tubes. Or Luis Gallo and his wineglasses. There's no such thing as old age in nature because, in nature, nobody grows to be old. Some species last longer because they're stronger and get eaten by fewer things. Sharks live longer than octopuses because nothing eats them. There's natural selection, and there are accidents and enemies. An accident is a storm or a very cold winter; a predator is an enemy ..."

"All the illnesses we call 'chronic'," I say, trying to get things straight in my head, "are old-*people* illnesses, that is to say, illnesses of the human species, who are the only ones, along with their pets, that get old."

"Right."

"When I get old, there aren't any glasses left."

"In short, natural selection quickly eliminates any illnesses that express themselves while creatures are still fertile, anything that implies an evolutionary disadvantage, because natural selection favours the ones who don't have them. If you're a wineglass that's bound to break within five years, better to give it your all the first time you get a chance to reproduce. It does you no good to survive, there's nothing for you there. If salmon held themselves back, that would be a mistake. For their genes, that would be a poor strategy. Better off going for broke and then dying, than returning to the sea and getting eaten by a shark, or than trying to swim upriver again, only to die without having spawned. There's nothing, Millás, it's just an atomic dance, regulated by the laws of matter."

"Okay, maestro."

"We can wrap it up in four points. One: there's no such thing as old age in nature, because individuals don't grow to be old. Two: natural selection isn't capable of dealing with what happens afterwards. Three: that's just the way it is. Four: so tailor your reproductive efforts to your life expectancy. If you're part of a species that lives to one hundred and forty, pace yourself. If you're only going to live to three, just go for it, go for it like in a hundred-metre race. Be semelparous."

He goes on, "And along with Medawar, write this other name down: Williams, George C. Williams. He was a biologist, and in the 1970s he published a study on longevity that we're going to be coming back to quite a lot. I'd just like to dwell on something he said that's very important for our book. If death is pre-programmed, if it comes about because of something internal, if it's intrinsic to the species, therein actually lies our salvation — nonsensical as that may seem. It's our salvation because there will only be a small number of genes — maybe even just one — that are responsible for this pre-programmed death. So, all we need to do is modify it, in order to become immortal. Then all we'd need to worry about would be extrinsic causes of death, like starvation, accidents, predators, war, pandemics, even meteorites. And if we manage to avoid all of those, we'd never die, we'd be like wineglasses that don't break because there's no risk of that ever happening. But if old age and death are due to the accumulation of multiple genes, as Medawar said, if the cause is this genetic burden, it may then be that we're lost, because we'd need to modify, one by one, all these genes that have been accumulating over time, since natural selection never encountered them in prehistory. And if there are lots of those genes, as seems likely, we're never going to attain immortality. Williams's conclusion is the following: there's nothing for it. You can choose to believe in death being pre-programmed, but Williams tells us not to hold our breath. So we need to work out whether he's right or not. Pre-programmed death, or genetic burden? That is the question."

I say, "I've made a note of all that, but tell me one last thing."

"What?"

"Had you agreed with the waiter that he'd break a glass just at the moment you were about to tell me about Medawar's test tubes?"

Arsuaga raises his eyebrows quizzically.

"And another question: did you ask them to serve us the octopus on enamelled iron and porcelain dishes, the 'lifetime guarantee' ones, so you could draw a comparison with the 'eternity' of the lobster?"

"Do you think I'm crazy, Millás?"

"No, I just think you're obsessed with staging."

Back home, going over my notes, I realise there's something missing, and I email the palaeontologist to say that we've talked about sex and about death, but not about altruism or cooperation. To which he replies: "Don't worry about it, there's still time to talk about altruism. As for sex and death, if you've understood it all, that just means I haven't explained myself properly."

FIVE

The grandmother hypothesis

Arsuaga called to tell me to buy a tracksuit.

"A tracksuit? What for?"

"I'm taking you to a gym," he said.

"And I can't go in my civvies?"

"You'll stand out," he told me.

I went to the sportswear section of a big department store close to my home and was wandering about in circles until a charitable shop assistant came over to lend a hand.

"I'm looking for a smart tracksuit," I said.

"A smart tracksuit!" she exclaimed, in disbelief. "What do you mean by 'a smart tracksuit'?"

I told her I was going to the gym.

"But only once," I added. "And I'd like to be able to go on using the tracksuit after that to go out for dinner."

She looked at me with pity then explained that the syntagm "smart tracksuit" was an oxymoron. Then, seeing my look of surprise, she explained that she was a philologist. She was actually the third or fourth philologist I'd found similarly out-of-place that month. What on earth is happening to philology, I wondered.

"There's no such thing as a smart tracksuit," the philologist concluded. "It's a contradiction in terms."

Once I had accepted this economic misfortune, we looked for a black one in a size that was surprisingly small — my diet, notwithstanding the excesses Arsuaga had forced upon me, was beginning to show results: I'd lost six kilos, especially around my waist. Buying clothes in size Small after months or years of getting them in Large is a mystical sort of experience. Standing at the changing-room mirror, looking at how well that lightest of outfits suited me, I felt like an Indian Brahmin. I thought all I had to do was sign up for some yoga classes — mental yoga, naturally — to find myself raised above my passions and attain, before he did, the ataraxy to which the palaeontologist aspired. I enjoyed the idea of telling him that I'd reached that state of mind, with all its tranquillity and a total absence of any desires or fears. He'd die of envy, even if he knew that envy isn't a very Epicurean feeling. When I told him on the phone that I had my tracksuit, he said we'd meet on 14 April at 9.00 a.m., outside his house.

"The anniversary of the Republic," I pointed out.

"It is," he said, "but our next outing has nothing to do with the Republic."

On the day in question, the moment the palaeontologist stepped out of his door and saw me in that black outfit that hung so well that it showed off my slim physique, he cursed. Seeing his surprise, I gave a wicked smile.

"I'm on a diet," I admitted before he had a chance to ask if I'd fallen sick.

"Please, Millás, don't turn into a slim dead person," was all he said.

Then, in the Nissan Juke, as we headed for a neighbourhood

in the outskirts called Montecarmelo, according to what I could see on the sat nav, he tried to sabotage my diet of five daily intakes, explaining that diet was a part of the niche in which one lives.

"A niche is your function," he added. "If you're granivorous, as partridges are, that means your nest is on the ground, and in turn that you're exposed to predators from the very moment you hatch. That means you're going to be nidifugous — you're going to leave the nest as quickly as you can — to evade those predators."

"And what's that got to do with my being on a diet?"

"What I mean is that the niche is either the full package, or it's nothing at all, and diet is just one more piece in the puzzle. But if you're happy with yours, I've got nothing to add."

"It's just that I'm on a diet that's niche-appropriate," I said in my defence, "which is to say, appropriate to a rather sedentary man who works basically with his brain."

"That's *you*, Millás, but your genes are Palaeolithic, and people from the Palaeolithic ate whenever they had the chance, with long periods of enforced fasting in between. Can you imagine a granivorous bird going to a nutritionist and being told to eat certain insects that don't exist in its environment? You can't live in one niche and eat things that are produced in a different niche. Fish don't eat mince."

"Did you not sleep well?" I asked, because he seemed to be irritated and needing to take it out on me.

"I slept fine," he claimed.

"So?"

"So, what?"

"So what the hell is the matter with you?"

"Almost the second I got up this morning," he admitted, "I

had to deal with a load of university bureaucracy. To name one thing."

"Bureaucracy is a part of life," I said in the voice of a Zen master, to show him how far he was from ataraxy, compared with me.

"I'll tell you something: people wear me out. Human beings, they're exhausting. Sometimes I just need to be on my own."

"I can get out if you like."

"Let's just drop it. What I'm trying to say is that you yourself may be good for nothing, but that isn't the case with your genes. Your genes are hunter-gatherer genes, they come from those who forded rivers, who crossed the tundra, who endured heat and cold, and who got to eat, if they were lucky, once a day. What have you done to deserve steamed hake?"

"That is actually what I had for dinner last night. How did you guess?"

"I just need to look at your face. But what have you done to deserve that fish?"

"Well," I argued, "today I got up at 6 a.m. and wrote an article for the paper."

"And did you do some cardio work?"

"I've walked a bit."

"That's not cardio, that doesn't get your heart going."

"I also had an argument with someone."

"I don't mean the kind of bump in your heart rate caused by an argument. One day, you're going to come and do a workout with me, then I'll buy you a Palaeolithic meal. Only if you want to, obviously."

———

We finally arrived in Montecarmelo, where Arsuaga left the car in the carpark of a tall building, all concrete and glass, which turned out to be a gym (GoFit) of colossal proportions, the big box store of gyms. I'd never seen such a thing. At the reception desk, a poster assured us in large letters that we would attain happiness through regular physical exercise carried out according to scientific methods. Technology and innovation, it concluded, thanks to which we would live longer and better.

We were met by the director of the establishment, since I've said on several occasions that the palaeontologist has friends everywhere, and she led us, via endless staircases and corridors, to a Pilates class that was about to begin.

The teacher's name was Marta Pérez, and every bit of her exuded health.

It was a class of about fifteen or twenty, though I should say more specifically fifteen or twenty *women*, as most of the students were female, ranging in age from fiftyish to seventyish. Each of them occupied her own space, marked out by a white square drawn on the black floor. Two of these spaces had been reserved for Arsuaga and me.

I noticed that my tracksuit was the most elegant, or possibly the newest. But everyone else, including Arsuaga, took theirs off, leaving only their shorts or leggings and their T-shirt. I couldn't take off mine, because I was only wearing my underwear beneath it. I cursed the palaeontologist for not having warned me. I felt a bit ridiculous.

Marta, the teacher, stepped onto the platform at the front of the class, and those glorious bodies, swathed in beautifully sculpting sportswear, all began to move. Arsuaga had told me more than once: "The human body is not a particularly lovely thing. But put it in leggings or neoprene, and we turn into gods."

Goddesses and gods, that's what those bipeds looked like to me, starting with the teacher, who was pure fibre. Nor did the palaeontologist — I had to admit — look at all bad in shorts and T-shirt, a beautiful and striking T-shirt, with an Atapuerca motif. Here and there, you could see the muscles he'd acquired on his long hikes across the Madrid mountains and on his constant visits to those prehistoric sites he was in charge of. Since we had taken off our shoes, I noticed he was also wearing designer socks, somewhere between grey and black, which it would have been a shame not to show off. Although I was slim like the rest of the students — thanks to my diet — I found myself rather out of place.

What happened next?

I have a vague recollection, nothing precise, because I was more concerned with what I looked like than with what was going on around me, but I do recall a few commands from the teacher: "Watch that sternum, shoulder blades down, find the resting surface on the floor! Visualise that surface, find the sticking point."

The goddesses and gods moved as one, as if in a ballet that had been rehearsed a thousand times before, while I tried to remember where my scapulae were, not to mention my sternum.

"Inhale and stretch," Marta went on, "one, two, three, exhale and open, one, two, three. And focus on a point on the floor, one, two, three."

The goddesses and gods raised their legs, bent, up and down. The voice in the background continued: "Sternum up, spine long, head nice and still, hips level, and … hold it there. Stretch, expand, but stick to the floor. When your sternum goes down, your waist drops, too. Ten, nine, eight, seven …"

The choreography was hypnotic. From time to time, I got

one of the specified movements right, and my muscles thanked me with a kind of electrical current that was them saying: we exist; we were dead and we have been revived.

Afterwards, at the sports centre cafe, sitting in front of some restorative juices that are a little sweet for my taste (and perhaps for my diet's taste), Arsuaga looks tired but content. His features have been restored to his usual levels of Epicureanism.

"Did you notice," he says, "that all those people knew anatomy? I got called into the dean's office after some students complained that I'd forced them to study it. They know less than their grandparents did. They don't even know where their glutes are."

"Right," I say.

"Have I ever told you, by the way, that the gluteus maximus — we've got three of them — is the largest muscle in the human body, and yet despite that, it has no known function?"

"I don't think so."

"Well, it would be good for us to cover that, because it has a bearing on sexual selection. If the glute, this enormous muscle, doesn't actually do anything — or at least nothing important — what the hell's it doing there?"

"It's being pleasing on the eye," I suggest.

"Now you're getting it, bravo. If you think about it, we're the only species with an arse. If gorillas had an arse, they'd look human to us. Give a chimpanzee a little waist and a nice arse, and it would seem sexy to us. But from a chimpanzee's point of view, it isn't at all. When they're on heat, their behinds swell up and turn red — the protuberance is enormous. For them, attractiveness lies in the perineum."

"It's so rewarding, that muscle, the perineum!" I exclaim from experience.

"It isn't a muscle, it's a region," Arsuaga corrects me.

"That might be even better. A mythical region, like Gabriel García Márquez's Macondo or Juan Rulfo's Comala."

"That is where," the palaeontologist continues, "the chimpanzee's attractiveness lies. Its seat is the most attractive part of its body, the part it uses to announce, literally to the four winds, using a visual and olfactory signal, that it's sexually receptive. That it's ready to copulate."

"Right."

"Aristotle said we are animals with an arse; given that, the arse has to serve some function. What function? To sit down on, was his deduction."

"Of course."

"But there's a small problem: it isn't true. You're sitting down now, yes or no?"

"Yes," I say.

"Introduce your hand between your behind and the chair and tell me what you touch."

"The bone!" I say in surprise.

"Indeed. So when you sit down, the glutes open out, in such a way that means they don't in fact, as Aristotle thought, act as a pillow. You are sitting on the ischium — that's the name for the bones you've got down there. When you stand up, the gluteus maximus covers it over, but when you sit down, it draws back, and as you've just seen, nothing gets left between your skin and your ischium."

"Which is why we put cushions on chairs."

"Don't talk to me about chairs. Chairs are, along with refined sugar, humanity's worst invention."

"Why's that?"

"Because the normal thing for human beings, when they get together with others to chat or to eat, is to get down on their haunches, without the buttocks actually touching the ground. 'Active rest', that's what it's called, because it creates a muscular tension that's very good for you. When your kids were little, didn't you find it hard making them sit down on the toilet to do a poo?"

"Yes," I say, remembering.

"That's because the normal position for defecation is also down on one's haunches. In the countries where they still do that, there's practically zero incidence of diverticula or haemorrhoids. If I could rewind history, I would eliminate refined sugar and the chair. The chair is the devil's work, believe me."

"I can believe it. But we were talking about the pleasingness of arses."

"Look, there are different theories about the function of the glutes: that they're there so that we can move from crouching to upright, so we can run long distances, or climb stairs ... But actually, none of those functions require that part of our body to be quite so bulky. So, if we put all of them aside, the only explanation left is that of sexual selection. The arse, the glutes, as you so elegantly put it, are there to be *pleasing*."

"Like the peacock's tail?"

"More or less."

The palaeontologist takes out his phone, goes online, and shows me an amazing amount of advertising for gym exercises whose only purpose is to obtain a great arse.

"Why do you think squats are recommended so much?" he goes on.

"To develop interesting buttocks."

"There you go. You'll also see why (whether you're a man or woman) it's so hard not to look when someone with a really great arse walks by. This is the sexual selection genes working at a distance. A gene of yours might bring about some alteration in you, but its effect, whether good or bad, stays within your body alone. But for the gene in a body that's ten metres away to give you a hard-on — that's pretty amazing, yes or no?"

"Yes."

"Right, so this shows the importance of sexual selection, which generally isn't given as much attention as natural selection. But this also brings us to something that has a bearing on old age and death."

"Go on," I say, opening up my notebook and threatening him with my biro.

"We've been in a class with older women who are in fantastic shape. The thing you saw them doing, when they sat down into a cross-legged position, and got up again — which they did with relative ease — is used as a fitness biomarker for firefighters."

"I can believe it!" I exclaim, as someone who failed to even embark on that particular exercise.

"In our species," Arsuaga continues, "all the statistics tell us that, from the age of forty-five, there's very little chance of conceiving, and from fifty onwards, it's almost nil. Which brings us to the phenomenon of the menopause, which is exclusively human. Well, with one exception: elephants also experience it. They live in groups, in families that make use of the grandparents' wisdom. All other species are fertile until the time of their death or, in any case, their reproductive functions deteriorate in parallel with all their other bodily systems. In short, the reproductive value of a forty-five-year-old woman is practically zero."

"And the man's?"

"Not zero, but it does decline in the same way, as he loses flexibility, respiratory capacity, et cetera. But with women, the end to their fertility is clearly pre-programmed. Which brings us to the question of whether death is pre-programmed."

"Tell me a couple of things that definitely *are* pre-programmed."

"The maturation process, for example: children are born, they grow, they reach sexual maturity, they suddenly shoot up in puberty … All of this is pre-programmed, it all depends on the programming. And the menopause, of course: with that, a genetic switch gets flipped, it happens abruptly, not gradually — there's no gentle downwards curve. The incredible thing is for women of fifty, sixty, or seventy, like the ones in that Pilates class, to be in such good shape at an age when they can't have children anymore. Why do you think we live on for such a long time after the reproductive function's done? Why don't we die at fifty?"

"You tell me."

"There's a theory known as the 'grandmother hypothesis' that explains it: the idea being that the menopause is justified by the child-rearing assistance provided by grandmothers in prehistory. In the Palaeolithic, as we've said before, anyone who managed to make it to adulthood would die at around seventy."

"That's the longevity of our species."

"Which mustn't be confused — this is a point I'll never tire of making — with life expectancy. After the age of seventy, the same happens with us as with pets: we prolong our lives by looking after ourselves, but it isn't natural life anymore, but rather assisted life."

"Right."

"If a woman had a child when she was sixty, it would end up being orphaned because the mother would die when the child got to the age of ten."

"But *why* is the menopause pre-programmed?"

"Because of exactly this, because having children when you're sixty would be a waste, an absurd investment of energy, when the child would most likely end up an orphan and die. Orphans, in the Palaeolithic — and not only in the Palaeolithic — had very little chance of surviving."

"But the community …"

"… yeah, and Pachamama, and solidarity among all the peoples of the world … There are statistics on it, so let's not argue," says the palaeontologist conclusively.

"Okay."

"It makes no sense for a woman to have children and then let them be orphaned. Whereas it makes all the sense in the world for her to look after her grandchildren, because while her children have half her genes, her grandchildren have a quarter — so two grandchildren are the equivalent of one child. In other words, a woman, at that age, would be far better off having grandchildren than children."

"How wise Mother Nature is!" I smirk. "And what about great-grandchildren?"

"Great-grandchildren," says Arsuaga, ignoring my sarcasm, "only have an eighth. The link, biologically and emotionally, gradually becomes diluted."

"And so the extending of life that goes from fifty to seventy we owe to grandchildren."

"In a sense. From fifty onwards, even if the menopause doesn't come into play, the reproductive value of a couple tends to be zero, because it's so likely that children produced after that will be orphaned at a very young age. And the lower our reproductive value, the less natural selection works to protect us against chronic illness. I'm talking, of course, on a species level. Nonetheless, we can still do *something* to aid the survival of our genes, and that is to look after our grandchildren. In other words, you shouldn't be wondering why we die, but rather why we live for so long."

"Some people complain," I say, "that grandparents, because it's so hard for their children to reconcile work and family, are too involved with their grandchildren."

"I laugh when I hear that, because it's really all biological. We live beyond the time we're fertile thanks to grandchildren. It's been happening since prehistory. 'The grandmother hypothesis', don't forget it, would be an excellent chapter title."

"Oh yes," I say. "Very good."

Arsuaga is silent a moment, staring into the void. Something is brewing inside his head. Finally, he says with a somewhat evil smile, "I know a place near here where they do really great bacon rashers. And it's the time of day for bacon rashers. Sound like a plan?"

"Not really," I say, "I'm on a diet."

"Well, I'll drop you near the Metro, and go have some bacon myself. A bit of time in my own company would do me good."

I try to look Zen about this, but it doesn't work.

SIX

Naked and Sated

By the end of May, I had lost eight kilos and I was five years younger. If not six.

That's how I felt.

There was a skip in my step, I leaped out of bed euphorically, wrote at dizzying speed. My old clothes fitted me. I retrieved four or five jackets from the loft that were as good as new but which I'd previously retired owing to the surplus kilos.

This is going to be my year, I said to myself in the mirror, but I was also the victim or beneficiary (who's to say?) of a bodily optimism I hadn't enjoyed in some time. The problem was where to stop, at what point to move on from losing kilos to focus on maintaining those I'd lost already, since when slimming reaches a certain speed (cruising speed, I suppose), there's a great temptation to go onto autopilot and just keep on racing towards the slimline. Being slim takes our inner mystic out of the equation. People on diets often make the same mistake as those trying to give up smoking: they start looking down on fat people and smokers. The voluntary loss of mass, in addition to altering our physical shape, also means trying to find spiritual vents for existential distress.

Meditation? Yoga? Pilates? Buddhism?

That was when I received an email from Arsuaga, which said the following:

> Dear Millás,
>
> The probability that a male of seventy-five, like you, won't be blowing out the candles on another birthday cake, is twenty-seven per thousand — that is, twenty-seven out of every thousand Spaniards your age die each year. It doesn't seem like much, but the problem is, there are a lot of dead bodies already on the pile, and it's getting bigger all the time. Next year, if you've managed to survive this one, the probability will be twenty-nine out of a thousand. Get to eighty, and it'll be forty-one in a thousand. At eighty-five, it goes up to seventy-one in a thousand. If you make it to ninety, it'll be one hundred and twenty-four in a thousand — that is, more than twelve per cent. Then you'll need to start worrying. The figures seem small, but this is like a race along an obstacle course, and the runners keep getting tripped up. It's why demographic charts of normal populations are pyramid-shaped, though not with nice straight walls, like you see in Egyptian pyramids. In Spain, the pyramid's inverted, of course, because no children are being born here, and at this rate it's going to end up more like a spinning top. All of which is just to say, while we're as healthy and good-looking as we currently are, we can rest easy … By the way, if you're still going with the diet, take care: if you look in the mirror and see

your skull, give it up immediately.

Abrazos,

Juan Luis Arsuaga

I looked at myself carefully in the mirror, to check whether I could make out my skull, and sometimes I saw it and sometimes I didn't. (Are teeth skull or aren't they?) That's what happens when you get obsessed with something: you're permanently confused.

In any case, should I put Arsuaga's eschatological email down to his sense of humour? Was it merely statistical information, or a veiled personal attack? It wasn't easy to tell. You never know with Arsuaga. He alternates between moments of Zen or Epicureanism and moments of misanthropy that really affect me emotionally, as they awaken my own aversion for humanity.

Aversion for humanity?

No, that's not quite it. What I had felt at another time, especially in my adolescence, though also in my youth, was a resentment arising from a misapprehension: the mistaken idea that the world owed me something. It's not good to live with that idea, which can become an obsession, eating away at you while you go on thinking — incorrectly — that it's eating away at everyone you hate. Through psychoanalysis, I had managed to eliminate it from my rational self, but possibly not from my emotional self. I know the world owes me nothing, yet still I feel it does. That same afternoon, when I went to see my analyst, after lying down on the couch, I once again raked up this subject that she and I had considered forgotten.

"I don't think I've managed to shake the idea I've got in my head that the world owes me something."

"From your head or from your heart?" asked the therapist, as though she had read my mind.

"From my heart," I replied. "I haven't managed to get it out of my heart. Of my head, I have: I know that, rationally speaking, it's stupid to think the world owes me anything."

"But it's lovely to think the world has some debt to us."

"Lovely and also effective, in its way," I replied.

"Effective how?"

"That imaginary debt is what made me write. Writing was a way for me to channel my resentment for that unpaid debt. Reading, too. Maybe I shouldn't say this, but I believe I read and write out of resentment."

"Would you stop reading and writing if it made your resentment go away?"

"Perhaps," I said, just thinking aloud. "Perhaps I'd replace the resentment with curiosity."

"But you have a reputation for being curious already."

"The curiosity is the screen, the mask for the resentment."

"And what's taken you back to this debt that the world supposedly owes you?"

Arsuaga hadn't yet shown up in my analysis, or if he had appeared at all, it was in passing, when commenting on events from our previous book. For reasons that are obscure, I refused to let the palaeontologist in there. I didn't want him invading this most private realm of my existence. I wasn't sure what to do. But finally, I let him in.

"I get the sense," I said, "that there's a sort of misanthropy in Arsuaga that I have repressed in myself. Which means that, far from curing it, I've denied it."

"Does Arsuaga feel the world owes him something? Do you think that?"

"I don't know."

"So ask him."

Which is what I did: I asked him via an email I sent that same night and to which, oddly — because it wasn't his style — he replied swiftly, in these words:

> No, Millás, I don't think the world owes me anything. For followers of Lucretius, the world/ self duality doesn't even exist. We are simply atoms that combine for a certain period of time, before separating in order to recombine once more. It's a privilege to be the combination of atoms we currently are — and which, by the way, are constantly getting renewed throughout our lives. On a personal level, I feel I've been blessed by the gods, I feel very lucky indeed.
>
> We need to carry on talking about the relationship between niche and diet.
>
> Abrazos

The world didn't owe Arsuaga a thing. It made me want to ask whether he'd answered with his head or his heart, but it seemed more sensible just to let things lie.

Soon afterwards, he summoned me to lunch at a restaurant on Calle Estébanez Calderón, right by Plaza de Castilla, which was called Naked & Sated.

Naked & Sated.

———

I asked myself whether even the name itself contained a hidden message, and I immediately answered: yes, it did, since they served Palaeolithic food. Not that they advertised it in those terms, but that was the purpose that lurked beneath the information I found online: "At Naked & Sated, we believe it's time to restore good sense and reset our body and mind. We want to put real food onto your plate, naked and fresh and in season, which will help you to take care of yourself and help us to take care of the environment."

Real food capable of resetting my mind. That's what I'd been looking for ever since I started losing weight: to connect to real reality (I often felt as though I was living in a simulated reality) and reset my mind. I soon discovered that losing kilos wasn't enough, hence my beginning to wonder about meditation, yoga, Pilates, Buddhism ... Could the Palaeolithic lifestyle be just what I'd been looking for all along?

The restaurant, with its kitchen in full view, turned out to be agreeably spacious — I don't know, maybe it was just that its spaces seemed both to open onto the outside (the street) as well as into the overall interior space. The decor, which I mentally classed as Californian, maybe with the odd hippy touch, combined to create an atmosphere propitious to great vibes, whatever "great vibes" might mean, or even whatever "Californian decor" might mean. All the same, and given my emotional state, I couldn't help but wonder whether it was a business or a temple I had stepped inside.

We took our seats at a separated-off area at the back of the establishment where we were being awaited by the man who would be our host, José Luis Llorente, better known as Joe Llorente.

"This gentleman," said Arsuaga, "is an ex–basketball player."

"Pleasure," I said, holding out my hand.

I was busy calculating how much further I'd have to go till I was as thin as Joe, and maybe till I had the calmness of spirit (or Epicureanism) he conveyed, when a waitress asked us what we wanted to drink.

"Wine," I said right away, automatically thinking ahead.

"I have nothing against drinking wine on special occasions," Llorente jumped in, "but I'd advise you to drink kombucha."

"I don't know what kombucha is," I said, uncertainly.

"It's a fermented tea," he explained, "that's very good for the microbiota."

I knew what the microbiota was, or the microbiome, because I'd done a couple of reports on the subject. Basically, the term relates to the group of beneficial bacteria that live all around our bodies, but specially in our digestive systems, and without which our digestion (though not our digestion alone) would be impossible. There are labs dedicated to the cultivation of these microorganisms that the food industry then introduces, for example, into yoghurt. When the packaging of a nutritional product says it includes "probiotics", it means it contains those kinds of bacteria we couldn't live without. Although it might look like we're their hosts, in practice, and only slightly exaggerating, you could say that it's them who are hosting us. If you've ever wondered about the meaning of life and haven't yet found an answer, don't rule out the possibility that our existence might be in the service of sustaining these microorganisms. A little while back, my digestion turned slow and heavy. I called a biochemist friend and told him my symptoms, and he told me that from seventy-five onwards our bodies stop producing the intestinal bacteria charged with keeping the inflammatory processes in check.

"So I would suggest," he added, "that you take a daily dose of *Lactobacillus gasseri* KS-13, as well as *Bifidobacterium bifidum* G9-1 and *Bifidobacterium longum* MM-2, to make up for these shortages."

I consulted my neighbourhood pharmacist and it turned out that that combination of microscopic creatures had been commercialised for some time under a brand of pharmacy products. I immediately got hold of a pack. These bacteria come freeze-dried, in capsule form, and they are revived in the damp environment inside the belly. I have to say that my gastrointestinal existence did a 180-degree turn with the ingestion of those bacteria, to which I am now addicted. It was then that I came to understand the growing acceptance of a treatment known as "faecal transplant", which consists of transferring the microbiome of a healthy patient to an ill one.

Well, if kombucha was good for the microbiome, nothing else needed to be said. I'm very keen on those strange lifeforms I have living inside me and to which my metabolism owes so much that I couldn't tell you whether I'm made for them or they for me.

To convey some sense of the quantity of microorganisms living inside us, keeping us in shape, Joe Llorente added, "We've got a hundred times more alien cells than cells of our own. We're a zoo."

However, before we went any further — stickler and stick-in-the-mud that I am — I carefully read the label on the drink and made a partly rhetorical objection: "But is this really healthier than wine? It says here it contains carbonated water, which I disagree with because it causes gas, and it also says that 'the sugar used is consumed in its totality during the fermentation process'. The word 'sugar' in any context at all puts me on my guard."

"But look, it's been consumed during the process of fermentation. You do see that?"

I shut up then because he was right: where there used to be sugar, there was now nothing; but when you believe that life owes you something, you distrust everything.

"What we're going to do today," Arsuaga said, to take the edge off the rather uncomfortable silence I'd created with my objections, "isn't exactly a meal, or even an 'experience', as the critics who hand out those Michelin stars say. This is part of a lifestyle."

And he turned to Joe Llorente and added, "I'm trying to explain to Millás the idea that diet is a part of the niche, and we need to bring back the best of what was an excellent lifestyle. It's to do with far more than just what we eat, in other words, because this lifestyle is trying to tell us that we're better off getting sugar from cranberries than from the sugar bowl."

"Why?" I asked.

"Because cranberries have fewer calories. Cranberries don't cause diabetes, or the terrible knock-on from diabetes, which is blindness. The sugar you get in the sugar bowl causes diabetes and real — as well as metaphorical — blindness."

"Right," I said, feeling unnerved.

"But even this wouldn't be the niche either," added the palaeontologist, his tone didactic, calm, Epicurean, sensual, sybaritic, voluptuous. "The niche would be the *getting* of the cranberries. A kilo of cranberries contains seven times fewer calories than a kilo of refined sugar, and you've also got to take away the amount of energy you'd have put into collecting them. The niche — as a totality — includes far more than healthy eating. It includes the exercise it takes to obtain the food."

"Of course," I agreed, seeing Joe Llorente's look of approval.

"The Paleo lifestyle," Arsuaga went on, "means far more than doing sport and eating well; it means living in harmony. You can't replace a walk in the hills or the countryside to collect cranberries by going on the treadmill. The sensations are completely different, because, as well as a body, we have a mind."

That often happens: just when I'm about to reproach the palaeontologist for his biologistic excesses, he makes some reference to the mind or the soul or the spirit and manages to disarm me.

I think he can read my mind.

"If you don't have the option to enjoy — body *and* soul — being in nature every day," he conceded finally, "well, *then* you can get on the treadmill. It isn't the same as the countryside, but at least it's better than sitting in an armchair."

Joe Llorente agreed again, with a nod, which clearly encouraged the palaeontologist.

"A Paleo lifestyle can include talking with your children, your grandchildren, taking some time for yourself … It's far more, anyway, than just eating well."

"What else?" I asked, so he could really run with the Epicureanism now coursing through him.

"Every night," he went on, "I hear a sound outside my house, something the neighbours seem to think is an alarm, because of its regularity. But it's a tawny owl, Millás."

"A tawny owl," I repeated, with a slight questioning tone.

"It's a nocturnal bird of prey, one that's very at home in the city," Arsuaga explained. "Being able to listen to the tawny owl without it bothering you — this also forms part of the Paleo lifestyle."

"But does it really bother people?"

"If you look online, you'll find thousands of people complaining to the city council. They want them taken away."

"What a shame!" I lied, since that night I'd slept with the window open and had repeatedly cursed the tawny owl, whose electronic-sounding song is identical to the beeping of my induction cooker when it goes crazy.

"It's a great bird to have around," Arsuaga insisted. "It eats all the cockroaches, for example."

There was a somewhat violent moment of silence, which was broken by Joe Llorente.

"I hadn't finished introducing myself," he said, in a tone of complaint, or almost. "I come from a line of sportspeople. We're fourth-generation now. My uncle was Paco Gento, the famous footballer. I've got three younger brothers, one of whom also plays pro basketball, and the other two played football for Real. My nephew, Marcos Llorente, is one of the top players at the moment. He's in the national team, and he won the league with Atlético. We're a particularly sporty family, but we're also particularly dedicated to health and diet. When I started out and I used to take my concoctions to training, people said: 'Where are you going with that birdseed?' They all laughed, but I ended up having the longest career of anyone in my generation."

"How old are you now?" I asked.

"Sixty-two."

As I've said, Joe Llorente is very thin, wiry, but his skull seemed quite prominent beneath the skin of his face, like when you're wearing very tight clothes and your body shape shows through.

"And you follow a kind of Paleo diet," I said.

"I do have that sort of life, yes, I'm our family's guru."

"Do you do the niche as a totality, as Arsuaga suggests?" I pressed him.

"I do. Part of our niche consists of being cold in winter and hot in summer. There's no heating in my house, or A/C. I take cold showers and sometimes I can't be bothered to get out because it's even colder outside the shower than in."

Just the thought of it made me shiver, which I hid by taking a swig of the kombucha. Then I went on with my questioning: "How much do you weigh?"

"Seventy-six kilos, five less than when I was playing."

"Perfect for your height," I said, as he was small despite having been a basketball player (he played point guard).

"I've lost muscle mass," he said sadly.

"Because you do less exercise," I tried to explain.

"Of course, I used to train five hours a day and now I train an hour and a half or two, and not every day. I go to the gym. With age, you lose a lot of muscle tone, and it gets harder to keep it."

"But you've got great biceps!" I exclaimed with genuine admiration.

"More or less," he acknowledged. "My legs are much better. I was going to come in shorts, but at the last minute I got embarrassed."

We all laughed, sipping at our kombucha.

"And so," I went on, "this Palaeolithic lifestyle isn't just an invention of Arsuaga's. There are those among us who imitate the life of the hunter-gatherer."

"That's right. I've got a Dutch guru, and I follow Arsuaga a lot, too, who I've been reading for years. I never miss a thing he publishes."

At that point, somebody came to take our order. We allowed ourselves to be guided by Llorente, who asked for a tuna tataki salad.

"But instead of chickpeas," he said, "we'll have sweet potato. And instead of rice, cassava. And bring a couple of buckwheat crackers with egg and avocado, too."

When the waitress left, Arsuaga told us he met a tribe of hunter-gatherers who lived in Tanzania, beside Lake Eyasi.

"The Hadza tribe," he added. "They're very fashionable just now. When I went to see them, twenty years ago, not even God would have visited them there. They speak a click language, like the Bushmen. They practise active rest, because they've never had the accursed chair to contend with. When you sit on a chair, the weight of your upper body and arms goes down through your arse. Your entire spinal column transfers its weight downwards. Resting passively like this is just the worst thing for us. The Hadza spend the majority of their time crouched down, one hand resting on someone else's shoulder, or on an object. Or unsupported. That means they're activating a whole load of different muscles — and you don't just have to take my word for it, it's been measured using devices that pick up on the electric currents, i.e., the muscular activity."

"Right," I said, staring at Arsuaga, who seemed once again to be reading my mind because he hurried to add: "Anyway, it isn't about taking these things completely literally. They're points of reference, let's put it like that. You can't turn a way of life into a religion, which is a risk all philosophies run — they sometimes get turned into religions, full of dos and don'ts. Diets are there for the sake of humans, not the other way around."

"My guru," said Joe Llorente, "teaches us that all this

business with fasting has to be done irregularly, and that you also need to enjoy it."

They brought us our food, which had a lot of green in it.

"I've asked for grasses," said Llorente, "because that's one of the deficiencies in our diet. Palaeolithic man stuffed himself with grasses just like dogs and cats do when they're out in the countryside. Grasses are good for our microbiota. We need to eat as much as we can and with as much variety as possible."

The salad of sweet potato, tomato, cassava, avocado, tuna, and sesame was divine.

"Avocados," said Llorente, "contain omega-3, which is very good for the brain and the joints, just like this tuna, which they've literally just seared for a second on either side. Red berries are antioxidants. The more colours you've got on your plate the better. Look, you can see they've put in goji berries, which are in fashion at the moment, and just a touch of black pepper, which irritates the intestinal barrier slightly, helping you absorb your food better."

"Good to piss that barrier off a bit," I said.

"To the right degree," replied Joe.

"Millás can't be a Palaeolithic man," remarked Arsuaga, "because he has to eat at 2 p.m. on the dot. If he isn't sitting down with food in front of him at exactly 2 p.m., he gets grumpy."

"That, like almost everything, is purely psychological," said Llorente confidently.

"You were saying," I jumped in, "that the important thing about the Paleo diet is that it's not regular, you don't eat the same thing at the same time every day."

"Yes," he replied tersely.

"Well, I've been doing a diet that involves eating five times a day and I've lost eight kilos. And I've got younger, too, or at least that's how it feels when I go for a walk."

"How much weight you've lost is irrelevant," said Llorente. "For Palaeolithic man, it's not good to eat five times a day, what's good is to eat five times *one day* and then once or not at all the next. Ana Maria Cuervo, who's co-director of the institute for ageing research at the Albert Einstein College of Medicine in New York, supports calorie restriction and irregular consumption so that the metabolic waste — which is toxic — isn't just left there to be re-digested. In other words, you've got to give your digestive system a rest."

"Ana Maria Cuervo," said Arsuaga, "knows what she's on about. She's a leading light. She's done exhaustive studies on the causes of Alzheimer's and all the other neurodegenerative illnesses; there's no one better on the whys and wherefores of ageing. The studies she's done on 'cellular cleanliness', or 'autophagy'— they're just stunning. Write this down: the cells in the body have their own cleaning mechanisms, which they use, as Joe says, to dispose of the waste produced by combustion. But they only do that when the cell is in a rest phase. If you spend the whole day eating, they never get a chance to rest, and the rubbish — or the metabolites, whatever you prefer — build up and up, to disastrous effect. The cells also do that cleaning while we're asleep, which means it's important to get a certain number of hours' sleep. I'm going to be interviewing Ana Maria Cuervo on my Radio Nacional slot."

"I like the word 'metabolite'," I said, gesturing for another bottle of kombucha.

"I said 'metabolite' to avoid saying 'crap'," explained Arsuaga genially, looking to lighten the scatological load.

"I still prefer 'metabolite'," I said.

"Cells need at least sixteen hours' break," added Joe Llorente, "to carry out the detox process."

"Does that mean," I asked, "that after this meal you won't have another bite to eat for sixteen hours?"

"I had dinner yesterday at 7.30 and I haven't eaten since, and it's nearly 3 in the afternoon now."

"You didn't have breakfast?"

"No, I've gone eighteen hours without eating, and sometimes I go a whole day."

Once we had finished the Palaeolithic salad, which had left me, at least, quite satisfied, Llorente said, "Now I'm going to order a free-range chicken burger. You guys up for that?"

We were.

"Over time," it occurred to me, "your body starts telling you what's good for you and what isn't. For me, for example, I like dairy, but it doesn't agree with me."

To which Llorente added, "There are some people, as Ana Maria Cuervo tells us, whose metabolic pathways allow them to eat whatever they want and they don't get fat."

"What's a 'metabolic pathway'?" I asked, fascinated by the expression.

"Good question," said Joe. "It refers to the way each person metabolises or digests."

"Pure biochemistry," the palaeontologist agreed. "The way your body breaks down the foods, which isn't the same for everyone."

At that point, just as the conversation was starting to turn very technical, they brought us the free-range chicken burger with a very multicoloured garnish. I asked Llorente to talk us through it.

"There are potatoes browned in olive oil," he said, "which

I will not be eating, but I'll be having every last morsel of that gorgeous sweet potato, and the tomato. Earlier, we had protein, with the fish and the egg, and there's more protein now with the free-range chicken. Don't forget: you've got to eat *free-range* animals. This here is mayonnaise and — holy shit! — they've actually given us cheese! Simple sugars aren't great for our intestine, but for that matter, neither are complex sugars."

"What's a simple sugar?" I asked.

"The kind you get in the supermarket, and the kind you get in refined grains, for example."

"And it's not a problem to mix carbs and animal protein?" I've got an oedipal prejudice about that, which comes from my mother's shellfish paella.

"That," said Arsuaga irritably, "is a complete myth, and isn't truly Palaeolithic either, because in the Palaeolithic, people would have eaten whatever they could get their hands on, not gone around being fussy. It sounds like Judaism — don't eat lamb together with milk, one of the Jewish taboos. Like I said before, it's this mania for turning everything into a religion. Look, Millás, it's one thing to lose weight, and another entirely to eat healthily. Everybody got really thin in Auschwitz. And hospitals are full of people who die when they're very thin indeed."

"My five-meal diet," I said, attempting to justify myself, "is as follows: in the morning, a piece of bread with cold cuts — ham, turkey breast, pork loin, et cetera. Midmorning, a piece of fruit. For lunch, a salad or vegetables with meat or fish. Midafternoon, another piece of fruit. At night, animal protein — steamed cod, octopus, or salmon, et cetera. It's not bad."

"The worst part, aside from failing to give the organism any rest whatsoever," said Arsuaga, "is your choice of breakfast: bread with cold cuts."

"It's the only time in the day when I have bread."

"If I were you, I'd try intermittent fasting," said Llorente.

"Millás's problem," said Arsuaga, turning to Llorente, "is that he confuses diet with niche. You need to make him an exercise plan, because he doesn't do anything."

"That's not true — I walk."

"Your kind of walking," insisted the palaeontologist cruelly, "does nothing from a cardio point of view. Your heart rate never hits sixty."

"So what you're telling me," I said, just giving in, "is that I should combine my diet with some specific exercises."

"The whole niche, that's the ideal thing," said Arsuaga. "Being properly Palaeolithic. Hunter-gatherers moved around a lot, they carried heavy weights, and every now and then they had to run at top speed, because there were predators around, too."

"Millás is seventy-five," Llorente pointed out charitably.

"I know, I know," Arsuaga said grudgingly, "but stress is good for everyone; the only problem is when it's permanent. What I think, ultimately, is that the most Palaeolithic thing of all about Paleo food is the conversation. To have a bit of time where you aren't rushing around, where you aren't looking at the clock."

"In that respect at least," I said, "I really am Palaeolithic."

"Yes, in *that* respect," said Arsuaga. "Now we've just got to set you up with an exercise regime that's right for your age."

"So in fact," I persisted, "my diet isn't as bad as all that, I just don't let my body rest and I ingest all my metabolites. That's also because I'm an introvert. My mother used to say, 'You're always so turned in on yourself; you ought to let it out.' Arsuaga," I added, turning to Llorente, "doesn't keep a single literal metabolite inside, but he doesn't keep the metaphorical ones to himself either. When he's got to something to get off his chest, he gets it off his chest."

"Is there dessert?" the palaeontologist interrupted me.

"There is," replied Llorente. "I've ordered a donut made with peanut butter, coconut oil, cassava, yeast, and dates. No added sugar. And they're going to bring us some chocolate, too — pure, no fat, which is very good for the brain. One hundred per cent cacao."

"Do you know what the five white poisons are?" Arsuaga asked me.

"No idea."

"Refined sugar, white flour, white rice, milk, and salt."

"In any case, after this meal, I'm not surprised you go seventeen hours without eating," I said, trying not to sound sarcastic.

"I think it's wonderful that Millás is giving his opinion," said Arsuaga. "Usually he's pretty inexpressive. He isn't the quickest when it comes to digesting new experiences."

"I've just got slow metabolic pathways," I said in my defence.

SEVEN

A question of size

I kept thinking about "cellular cleanliness" or "autophagy", and a few days later, I listened to the palaeontologist's interview with Ana Maria Cuervo on Radio Nacional. She was comparing the human cell (and we have thirty-seven billion of them) with a home you need to clean regularly so the rubbish doesn't end up consuming its inhabitants. It made me think of that day each week, in my childhood, when the whole house would be turned upside-down so that not a single speck of dust would be left anywhere. On those days (Fridays), if we didn't have school, as happened during the summer, we kids would be chucked out into the street to clear the area and we wouldn't be allowed back in until every room was clean as a whistle. Woe betide anyone who stepped on a recently mopped floor or stained a recently disinfected toilet! The house, like a cell, needed to be unoccupied so that the periodic maintenance could be carried out.

Ana Maria Cuervo, who works in the US, was lamenting the fact that people in that country eat so much (and so badly), since their cells never get the chance to dispose of the waste, which certainly accelerated the ageing of the cells and, as a result, that of the people, too.

I had often thought about my own ageing, but never about the ageing of my thirty-seven billion cells, which I started to think of as thirty-seven billion bathrooms of so many messy bachelors. A society of morbidly obese people could be very clean on the outside, but pure Diogenes Syndrome in its constituent parts.

I called the palaeontologist.

"Hey, is there any way of knowing what state our cells are in, if they're clean or dirty, old or young?"

"There is."

"So let's find out."

"One thing at a time," he said, "all in due course. We'll move on from big to small in a couple of activities I'm planning. For now, though, I'd like you to come with me tomorrow to the Museum of Natural Sciences."

"What for?"

"One goes to the Museum of Natural Sciences in order to go to the Museum of Natural Sciences," he replied.

"Okay," I said.

It's the following afternoon, and Arsuaga and I are inside the Museum of Natural Sciences, standing in front of a dissected elephant that's about five metres tall. It's like a building.

"Has this got something to do with that big and small thing you told me yesterday?" I ask him, disappointed, obsessed as I am with cell cleanliness or autophagy.

"No, it's nothing to do with that."

"It's almost as though they built the museum around the elephant," I say, "because if you look, you'll see it doesn't fit through any of the doors around here."

Arsuaga glances around, then looks surprised.

"You're right," he says.

I'm not one of those people who's desperate to be right all the time, but it does reassure me that my rightness gets acknowledged every now and then.

"At any rate," he adds, "we aren't here to compare the doors and the animal for size. Look at his scrotum."

I do look at his scrotum, which is at nose height for us.

"Have we come to see the elephant's scrotum?" I ask.

"It's of particular interest, because of what I'm about to explain to you. And stop looking for big meanings in everything. This is nothing but an anecdote, but it is interesting."

"Okay."

"The elephant belonged to the Duke of Alba. It was assembled by the Benedito brothers, who were the best taxidermists in the world, and transported along the Castellana on an enormous truck. Now here's the good bit. Take another look at its scrotum."

"It's pretty hard not to."

"Except that elephants don't actually have scrotums, because their testicles are held inside the abdominal cavity."

"And how did the Benedito brothers get this wrong?"

"It simply never occurred to them, and they failed to consult any manuals. So this is how it ended up, poor thing, to be seen like this by all the children and adults who visit the museum."

"Seems like a pretty amazing mistake to me!"

"Nobody's perfect. By the way, the physiological question of an elephant's internal testicles is a problem that's still to be resolved, because for the sperm to be reproduced, the testicles have to be at a lower temperature than the body. Ours, of course, are on the outside."

"I'd noticed," I say.

"Cetaceans, the whale family, have them inside their bodies. Now, that makes sense if a creature lives in cold seas, but we find the same problem with those that live in warm seas, the same conundrum as with the elephant: we know nothing about the mechanism by which they cool them. It may be some kind of refrigeration by the blood."

The palaeontologist and I are silent a few moments, in profound contemplation of the refrigeration system of elephant testicles and those of cetaceans that live in warm seas, though also — at least on my part — about the refrigeration system for human testicles, clothed in the dual textile covering of underwear and jeans.

"But what I was wanting to get across to you," says the palaeontologist at last, turning around and pulling a small book from his pocket, "is to do with life expectancy, which is such a concern for us, and with longevity."

"Let's do it." I ready myself to take notes.

"This is *Natural History* by the Comte de Buffon, a French scholar from the eighteenth century. This volume here — there are quite a few, written over a fifty-year period — is dedicated to the human being. And although it's full of mistakes, there are also a lot of things he got right, and I'm forever consulting it. He touches on the life expectancy of elephants, hence why it's useful to us now. Aristotle had said that large animals live longer than small ones, that is, that life expectancy is to do with size. We'll see if that's right or not. Buffon observed that small animals have an accelerated life, a cardiac frequency greater than that of large animals. The higher the body temperature, the greater the number of beats per minute."

"In short," I conclude, "they live quickly compared to bigger ones."

"But Buffon doesn't put the shorter lives of small animals down to this, as later scientists did. What distinguishes larger animals is a lower basal metabolism."

"What's basal metabolism?"

"Didn't you do natural sciences in secondary school?"

"I've forgotten."

"Basal metabolism, which comes from the fact it's the lowest or base level, is the minimum energetic consumption we require merely to live. If we compare species of different sizes, like blue whales and shrews, the figure has to be divided by body weight, obviously."

"Right."

"Or put another way: the calories needed for our organs to function in a rest state."

"The ones I use up just watching TV on the sofa, for example."

"Exactly," says the palaeontologist. "Right, so the basal metabolism of large animals like elephants is lower. Buffon observed that small animals, as Aristotle pointed out, live shorter lives than large ones, and that they live accelerated lives, but he didn't take it any further. However, might this hold the reason for ageing?"

"I don't know," I reply, looking up from my notebook, because the question does seem to be addressed to me. "But in any case, you look at the face of a mouse and it looks like it's always worried, like it's constantly beset by a thousand dangers. An elephant's is more relaxed. Worry kills, I can tell you that from experience."

"And yet," Arsuaga continues, "remember that the number of heartbeats in a mouse's life and in an elephant's life is the same, only elephant heartbeats go slower, more leisurely."

"That's what I was trying to tell you: permanent anxiety, unease ..."

The palaeontologist, oblivious to my comments, resumes his reflections: "It's as though there were a fixed amount of life, constant across all mammals, and they just parcel it out in their own ways. Some mammals fritter it away in just two years, and others make it last eighty."

Saver mammals & spendthrift mammals, I write in my book, to help myself understand.

"Recent thinking," Arsuaga goes on, "and make sure you get this down, because it's important, and because everything's ultimately linked to food — has explained this difference in life expectancy in terms of *oxidative stress*, which is what has led to the boom in the antioxidant industry."

"I take melatonin."

"But what actually is oxidative stress?"

"I'm not sure."

"It's the accumulation of the by-products of oxidation. If you live an accelerated life, then you get a build-up of these toxins, these metabolites or waste products from cellular combustion — everything we were discussing the other day when we talked about cellular cleanliness, or autophagy. And a mouse accumulates them in a shorter time than an elephant does because its basal metabolism is far higher than an elephant's."

"Which brings us back," I say, "to the subject of the Palaeolithic diet or intermittent fasting. That is, to the advisability of allowing cells to rest so that they can get on with the task of cleaning up all that rubbish."

I'm alarmed by the idea that the palaeontologist has dragged me here just to interfere yet again with my diet, which is still bringing me considerable satisfaction.

"Now," continues Arsuaga, "we're going to add a touch of maths to the story, let's see how you get on with this."

"Badly, I can tell you now, it'll be badly," I say.

"Ultimately, if you think about it, the answer lies in the surface-to-mass *ratio*. The larger the animal, the lower the ratio of skin surface to its overall mass. Do you follow?"

"No."

"In other words, the mouse, relatively speaking, has more skin than the elephant."

"Okay, now I do. It's always possible to put things more simply."

"But where do we give off heat from?" asks Arsuaga.

"From our skin."

"Exactly. Meaning that the animal with more skin in relation to its mass will lose more heat, and its metabolism will be more accelerated. A small animal loses a lot of body heat because, ratio-wise, it has a lot of skin. Elephants, on the other hand, since they have less skin — again, in relation to their mass — retain the heat better, meaning their energy consumption is lower."

"I understand that."

"Now I'm going to use a mathematical formula to explain it to you, so that you aren't just accepting it as a piece of dogma."

"I'm not accepting it as a piece of dogma. I'm accepting it because I trust you, and you're a scientific authority."

"Well, you shouldn't. Now, write down this formula, please."

"I don't think I'll be including the formula."

"Why?"

"Because books with equations sell fewer copies. You remember *A Brief History of Time*, the famous book by Stephen Hawking?"

"Yes, it was a bestseller."

"Well, when he first wrote it, it was full of equations. His editor sent the manuscript back to him, saying that each equation would mean a hundred thousand fewer copies sold, and publishers know what they're talking about when they talk about love. Hawking, who needed the money, took out all the equations and the book was a global success, even if people didn't understand a word of it."

"It seems slightly cynical as an argument."

"Not to me. I want to write a book that people understand and that people buy. The way you've explained it so far, people will understand. But start sticking equations in now, and we'll mess the whole thing up."

Arsuaga is silent a few moments. It doesn't look like he's going to drop his mathematical formula.

"And what does it matter whether it sells or not?" he says. "Do we really need the money?"

"Speak for yourself," I say.

"At least put the equation in a footnote," he suggests.

"I hate footnotes, which scientists seem to love so much. If it's important, don't just put it in a footnote. If it's not important, cut it."

"Is that your final word?"

"It is, I'm not prepared to budge on this."

The palaeontologist and I size each other up for a moment.

"It's normally you who wins all our battles," I add, in a conciliatory tone. "Go on, let me have this one."

"Just tell me you'll think about it," he says finally. "Right, we've dealt with the elephant thing, but look at this beautiful giraffe over here."

And indeed, only a few metres away, there stands a giraffe

who also — it seems to me — would have struggled to get through any of the doors around us.

"Another large mammal," I say.

"Which is therefore destined to have a long life. The important thing here is that Lamarck used a giraffe in his explanation of evolution."

"You told me once," I recall, "that I was afflicted with Lamarckism and that you were going to cure me."

"That's where we're headed. According to Lamarck, if, for example, you use your arms a lot to take things down from the top shelves in a bookshop, they'll end up becoming longer, like giraffes' necks, which can carry their heads to the uppermost parts of certain trees. Lamarck thought the information would go from the body to the genes, when in fact it's the other way around: it goes from genes to body. Lamarckism has become extremely widespread, because it's intuitive — but science, as we never tire of saying, is *counter*intuitive. However much you stretch your arms to get things down from the top shelves, they aren't going to get any longer."

"So how would they get longer, then?"

"If — and this is just an example — long arms became sexually attractive, and people with long arms started to be selected as mates by people who also had long arms. If in this selection process, individuals with the longest arms were constantly being favoured, a moment would come when your descendants would be able to reach a book you can't. The information, as you can see, goes from the inside — from the genes — to the outside."

"Right, that's the same explanation you gave me for the shape of Japanese people's eyes."

"It's called 'sexual selection'. And I'll just take this

opportunity to say that, nowadays, the length of a giraffe's neck is thought to be the result of various things, but perhaps the most important, or one of the most important, is yet again to do with reproduction, because they fight, they bash one another using those long necks of theirs. Anyway, for them, the neck is for fighting with."

"Lamarck," I say, "was intuitive."

"Oh, everyone's Lamarckian — I often meet doctors who reason like that."

"Well, I reckon I've been cured of my Lamarckism. Thank you very much. So shall we go now?" I add glancing at my watch, as I'm in a bit of a hurry. Besides, the museum guards have told us off a couple of times because we haven't been abiding by the route marked with arrows on the floor. I can't bear conflicts with authority.

"No, there's something else."

"What's that?"

"The giraffe, with that long neck, has seven cervical vertebrae. All us mammals have seven cervical vertebrae, regardless of how long our necks are. Go on, count them."

I bring my index finger to the back of my neck, but I can't really distinguish them properly. Fortunately, we do have an okapi skeleton to hand, and also a whale's. The whale's, which is vast, is hanging from the ceiling. They both also have seven vertebrae.

"Whereas," says Arsuaga, "with birds and reptiles, how long their necks are depends on the number of cervical vertebrae they have. A swan has far more than a duck."

"And so," I summarise, to speed things along, because I've got an appointment with my psychoanalyst, "only us mammals have a fixed quantity of these vertebrae, irrespective of the length of our necks. The only thing we can do is lengthen them or shorten them."

"Right, and this is important. Why?"

"I don't know," I reply impatiently, checking the time again.

"Because it tells us a lot about evolution in nature. All animals, including us — but not including sponges, coral, jellyfish, and a few others — are bilaterally symmetrical. We're made up of two practically identical halves joined down the middle. Now, what modern genetics has discovered is that all bilateral animals are constructed in a modular fashion. This is a very clever trick on the part of evolution. What happens is that each one of these modules then becomes modified and specialised. But if you take any bilateral animal, its body is composed of this aggregation of repeating modules."

"Like those houses — which, precisely, they call *modular* — that are made up of blocks of concrete and which can be made bigger or smaller depending on the modules you add or remove?"

"More or less. The best way to build complex forms is by using modular structures. You build a train ..."

"... and then you link up the carriages," I complete his sentence with another glance at my watch.

"The first module has the engine in it; the second is the sleeper carriage; the third is the restaurant car; the fourth, whatever you want. Once you've got a modular system, you can go along specialising each of the different modules: you put legs on one; wings on another; another is the abdomen; another, the skull, onto which you add the jaws ..."

"Very good," I say.

"Well, a living creature invented this six hundred or seven hundred million years ago."

I'm about to tell him that I've got to go, but now this modular business has started to reel me in.

"And what kind of living creature was that?" I ask.

"A kind of worm," he replies. "Up until then, you didn't have any animals that were organised in a modular way — think of sponges and jellyfish. But then an animal came along, between six hundred and seven hundred million years ago, and constructed a living system with a modular design, and that's where we all come from. This goes for mammals, birds, flies, for everything, with the rare exceptions I've already mentioned. The modular system is controlled by a set of genes called 'hox', from 'homeobox'. Homoeotics, that's the repeating parts."

"They're like boxes?"

"Boxes — exactly. The genes that control the unfolding of the modular system that comprises us — us and fruit flies alike — are called 'hox genes', and they're the same in all bilateral animals. We're all made the same, on the basis of a system of modules that were originally the same but that then, through the different evolutionary lines, have gradually specialised. Which means that what works for the fruit fly also works for you. This is a very modern idea, from the 1980s and 90s, it hadn't come in when I was at university."

"And," I venture, "we're made up of boxes: the cranial cavity is a box, the thoracic box, the abdominal …"

"Modules, they're all modules. And for some strange reason, the hox system, in mammals, prevents any more than seven cervical vertebrae from forming. Nobody knows why, but it could be to do with another concept that is important for our book: pleiotropy."

The palaeontologist has just introduced a new term. I'm obviously going to have to give up on my therapy.

"You seem on edge," he says. "Is something wrong? Is it because of the equation?"

"No, I've forgotten all about that."

"So what is it?"

I hesitate, but then finally confess.

"It's just, I had an appointment with my analyst, but I'm not going to make it now."

"You're having psychoanalysis?"

"Sure."

"And do you talk to your analyst about me?"

"You're not that central to my life."

"Well, ditto, but if I was having psychoanalysis, I'm sure you'd have come up at some point. We've written a book together, for goodness sake! And we're doing another one."

"Okay, maybe I'll mention you in my next session."

"Thanks. So can we get on with pleiotropy?"

"Let's."

"So, this is the name we give to a phenomenon whereby one gene can produce more than a single effect. Contrary to what people believe, genes actually encode various things simultaneously; they're related to various features. In the mammalian hox system, it appears that the number of cervical vertebrae must always stay the same due to the fact that certain individuals, including humans, developed an extra vertebra and either died young or failed to reproduce."

"Having an extra vertebra, especially in giraffes, might look advantageous at first sight."

"But it isn't. That's why they have to be kept in check. Seven per individual, full stop."

"And does this," I ask, "have anything to do with longevity?"

"This has to do with the way evolution functions. To put it

simply, I want to know why we die at ninety, and mice at the age of three. That's what concerns me."

"Me, too."

"Well, then you need to learn to read between the lines of everything we've been looking at, and get ready to read between the lines of everything else we're going to move onto next."

"Why between the lines?"

"Because you're clever."

"Thank you."

We continue our walk through the museum, which is empty at this time of the day, under the alert gaze of the guards, who keep drawing our attention to the arrows on the floor marking the conventional, compulsory route. These arrows remind me of the ones they put in giant IKEA stores, partly so you don't get lost, partly so you're sure to buy something in all the sections.

Suddenly we stumble upon an enormous tank in which a giant squid spends its days, or maybe its non-days, because it's dead. Despite the fact that all the other animals in the museum are dead, too, this corpse is especially striking, as though we've found ourselves before a fellow human. The museum has just turned into a morgue.

"Is it floating in formaldehyde?" I ask, as a way of diverting my initial thrill towards something more technical.

"I don't think it's formaldehyde," says Arsuaga, "but it's some sort of preservative."

"Well, it's been excellently preserved, it looks like it only died yesterday."

"This is the famous *Architeuthis* squid — it's the animal with the biggest eyes in the world."

They aren't just big, they're also beautiful and deep and filled with meaning, so that staring at them is like staring into the abyss, though I don't know whether it's the abyss of evolution or of one's own consciousness. Casting our everyday gaze upon these uncommon eyeballs is quite bewildering.

"And those," adds Arsuaga, still referring to the eyes, "aren't the original eyes, obviously, but the taxidermists did a good job of copying them with the use of a lightbulb."

"How many watts?" I ask.

"What does it matter how many watts?"

"I don't know, it's just that I find them touching, as if they were real."

"These animals are two metres long," continues Arsuaga, "seven if you include the tentacles, but there are even bigger ones out there. It's the longest invertebrate in the world. They inhabit the deep seas, and when they reach two or three years old, they reproduce and die. Such a huge unfurling of life, all just to reproduce once and die."

"What a high energy cost," I exclaim, "in the service of what — from a capitalist perspective — produces such small yield!"

"This one," Arsuaga specifies, "came up in fishing nets off Malaga. It was brought on board alive, but quickly died."

We say goodbye to the giant corpse, in which I believe I have discerned a kind of reserve common to all beings that are truly redundant, and obeying the floor arrows, we proceed along the circuit without being reprimanded by the authorities. Arsuaga wants to show me something nearby, but which, thanks to the compulsory circuit, is far away.

"This," I say, thinking aloud, "is like if you went to the Prado just to see *Las Meninas* and they made you walk through every single room."

Finally, we arrive at a vast dissected Galapagos tortoise.

"These live to over one hundred and fifty, sometimes over one hundred and seventy," says Arsuaga.

"And do they reproduce much?"

"They do what they can. But take a closer look, and think on this: what do you and I do over and over again?"

"I'm not sure what we do. You tell me."

"We ask ourselves why certain species live to eighty, others to one hundred and seventy, and why others, like the squid we've just seen, barely make it as far as three. That's the underlying question. My little girl used to say to me: 'You just need to study turtles, they must have some molecule that could be taken as a pill, and that'll be an end to it.'"

"That's what antioxidants claim to do, that's why I take melatonin," I say.

"I'm sorry to have to tell you that the antioxidants aren't working."

"Well, they're working for the industry. They're making them millions."

"But they don't work."

"The companies become richer and they live longer," I insist.

"They work for the companies, but not for you."

"Fine: you've said that's the big question. But you haven't given me the answer."

"You wouldn't be hoping for a Nobel Prize in Physiology or Medicine, would you?"

"You reckon they could give you and me a Nobel for this book, then?"

"For Medicine, yes," Arsuaga decides.

"Physiology or Chemistry, it's all the same to me, but for this book."

"The one for Medicine would be better," the palaeontologist

continues in his delusion.

"I'd prefer to get Medicine than Literature," I say, just musing.

"If we manage to work out why certain species live to the age of three and others to eighty, whether it's down to death being pre-programmed — and we light upon the little bastard of a gene responsible for the programming — or because over the years you get a build-up of a toxic substance in the cells, and that was something we could somehow eliminate …"

"… they'd give us the Nobel," I conclude.

"It would mean we could all live as long as we liked."

"Let's give it a try."

Arsuaga abandons his dreaming and points out that, almost without realising it, we have arrived at the final floor of the museum.

"In here," he says, "there's a research area with a fantastic collection of insects, but it isn't open to the public."

But then a door opens and a friend of the palaeontologist's appears (a Sapiens knows people everywhere, as I've said), and she invites us to come through. Her name's Mercedes París, and she's worked for many years in the management and care of the entomology collection, which is superb.

"We only conserve and keep six-legged specimens," she says. "Spiders and the rest are handled by another conservator. We have a problem here, which is that everything you can see is dead matter and there are creatures that feed on dead matter and they could wipe out the whole collection. If you're not careful, one day you'll find a box with nothing on its pins and a just bit of dust at the bottom. That would be game over."

"And how do you protect the collections from the corpse-eaters?" I ask.

"We *used* to use insecticide —" she says.

"Insecticide protected the dead insects?"

"Yes, it's a bit of a paradox. In practice, what it stopped was the dead insects getting eaten by living insects. But insecticides aren't healthy. Now we have a system of prevention that involves each new specimen spending more than two weeks in a freezer at minus forty degrees. It's assumed that if it's carrying any bugs, they'll die. Besides, in the rooms themselves, as you can see, it isn't hot. It never gets over seventeen degrees. It's tricky in the summer, but even if some little grub is still alive, it's not very active. We have almost two hundred cabinets, and each one fits one hundred and twenty boxes."

As she talks, she shows us boxes that are like little portable display cases for dragonflies, beetles, butterflies, with each specimen more beautiful than the next. They look more like the work of a brilliant silversmith than of nature. We also see grasshoppers whose wings are identical to the leaves of the trees on which they come to rest.

"This mimesis," says Arsuaga, "is the proof that Lamarck got it wrong. Only natural selection could do this. How could an insect make itself look like a leaf? By putting on a leafy expression? Not possible."

"I'm getting less and less Lamarckian," I admit.

We see a giant stick insect and a group of butterflies whose wings look like rose leaves. Then I recall an article about mimesis that I read when I was young in the *Enciclopedia Espasa*, which said that some caterpillars assume the shape of bird excrement to avoid those same birds eating them. I wondered then, and I wonder now, if it's worth keeping your life if, in exchange, you've got to look like a piece of shit. I still have an ambivalent attitude towards mimesis. On the one hand, it does fascinate me; on the other, it seems like one of nature's most humiliating tricks.

There is another creature that, in order to defend itself from its predators, takes on the appearance of a dead body covered in mould. Sure, they're safe, but at what cost? "Don't stand out, son," that's what mothers used to say in my day when they saw us going out with our beards and duffel coats. Not standing out meant going unnoticed, or being taken for a turd or a decomposing corpse. The important thing was not to be looked at, because once you'd been spotted, you could become food for species that are more violent than yours. There was one sergeant in military service who gave the same advice as our mothers, though more expressively: "Anyone fat should get thinner, and anyone thin should get fatter."

Don't look like an Arab, in short; or black, or Chinese, or like an anarchist, or even a social democrat. Don't stand out. Camouflage yourself. Get thinner if you're fat and fatter if you're thin. You haven't chosen the best moment to be different, kid, what are you trying to do? Try not to look this or that, neither meat nor fish. Hide your ideas, don't dissent, don't take sides, don't stand out. If an insect doesn't see anything wrong with being mistaken for a dry leaf, why all this fuss about looking like somebody? Act like you're going down the stairs when you're going up, and like you're going up them when you're going down. Don't raise your voice, keep up appearances, get thin, get fat, come, go, go out, come in. In short, *survive*, pretend to be a turd, a stick, a piece of bark. And home before ten.

That's what I thought when I read that article about mimesis, which, when seen in the boxes at the museum, turned out to be so fascinating and so artistic.

It was already getting late when we left. My psychoanalyst was going to charge me for the session, as I hadn't cancelled in time.

EIGHT

Kilos that come and go

"Are you starting to get an idea of what old age is?" Arsuaga asked me on the phone.

"Old age is a country," I said.

"A nice country?"

"I'm not sure. I'm still a stranger here."

In strictly chronological terms, I'd been old for a few years already, but I hadn't been aware of this until the palaeontologist and I started looking at the subject. I suppose this obliviousness came from the fact that I'd never stopped working, ever since I was very young. Even now, at seventy-five, I work more than I did at forty, because things just happen and also because I've lost my fear of writing. To be more precise, I've lost my panic. The panic used to paralyse me; whereas fear I actually find stimulating. I don't have such a tense relationship with syntax as I used to. There are even days when syntax acts like a friend and flows almost completely smoothly. It's also true that, on those days, I am suspicious of its intentions. Writing consists of bringing what the words want to say and what *you* want to say into agreement. If you allow words to speak for themselves, they will say nothing of interest. If you give free rein to your verbal ravings, what you write will fall flat. A literary text is the result of this negotiation, and it's sometimes an exceptionally hard one. Beware of syntax when it appears open to dialogue!

On the other hand, ageing is not a linear process, or not entirely. It's true that ultimately everybody dies, but sometimes they take three steps forward and two steps back. I was better between sixty and seventy than between fifty and sixty. In better health, a better state of mind, more creative … You can head towards death on one of those toll highways, going at one hundred and twenty kilometres an hour, or you can use the byways, with all their detours, never going above eighty, constantly stopping to take in the unrepeatable scenery.

This lack of linearity, combined with the healthy stress of work, had stopped me, in short, from noticing that I was old. When I did, I also understood how this condition, that of old age, was like a country. I am now attempting to ascertain who it is that governs here, what the local customs are, what language the inhabitants speak, et cetera.

"Well," concluded Arsuaga, "you're old already, and I'm on my way there. I hope you'll have done some more reconnoitring by the time I catch you up."

"I'm on it," I assured him.

"But I called," he went on, "to tell you I've managed to get us an appointment with Juan Antonio Corbalán."

Corbalán is a renowned Spanish cardiologist who in his day was point guard for the Real Madrid basketball team as well as the Spanish national side, with whom he won more than one hundred and twenty caps, playing many games alongside our old friend Joe Llorente. He works at the Vithas International medical centre in Madrid, on Calle de Arturo Soria, in whose sports-medicine unit he receives us one Friday in early June ready to give us — both the palaeontologist and me — a full check-up, from which he will ascertain our respective decline or vigour, relative to our ages.

There were things about myself I preferred not to know, but Arsuaga insisted we couldn't make a proper examination of old age without observing it in ourselves.

"And if we don't like what we see?" I asked.

"We'll face up to it, like the adults we are," he replied.

Doctor Corbalán had the natural elegance of those doctors on American TV shows. The white coat suited him so well, like it was invented just for him. His genial face, his tact, his whole manner, in short, conveyed a sense that it would be impossible for him to ever be the bearer of bad tidings. Or at least that he'd impart them to you in such a way that they wouldn't hurt. His office walls were white, the blinds on the large windows grey. Everything oscillated between these celestial colours. The light was milky, though not sickeningly so, and seemed to come from everywhere and nowhere at once. I have thought about that a lot, about the lighting and the paint that could, theoretically, have seemed aggressive, but which in practice conveyed a powerful sense of physical and mental cleanliness. You felt at home between those four walls, where, besides the doctor's desk, there were various pieces of equipment, also white or grey, among which I could make out a treadmill, for stress tests, and what looked to me like an EKG.

The normal thing is for doctors to see their patients individually and in private, but Arsuaga and I presented ourselves as though we were just one person, or as if we were conjoined twins or whatever, such that he got to witness my tests, and I his. Later I would wonder where this renunciation of our right to privacy had come from, whether it had come from him or from me, though it was undoubtedly the result of a tacit agreement. I think we like competing. We did care, of course, how each of us was doing, but maybe we cared especially about how each was doing in relation to the other. I was glad I had lost nine kilos, which I suspected would count in my favour, if not in all then at least in some of the tests we would undergo.

The first surprise was that in recent months Arsuaga had gained exactly the same number of kilos as I'd lost. He said it himself, looking at me incredulously from the scales: "I've put on your kilos!"

"They had to go somewhere," I replied, thinking pettily that I'd just had something of a win.

I volunteered to be the first to get examined by the doctor, so I took off my shoes and stripped from the waist up, under the gaze of the palaeontologist, who seemed a little worried.

"You've come in your tracksuit," he observed.

"I have," I replied.

"It'll end up paying for itself after all," he said.

From that point on, I began a battery of tests that I tried to memorise, though later, on trying to recall them, they all got mixed up. I do know that after Corbalán had made a note of my weight, my height, and my age, I sat down in a chair from which he quickly asked me to get up again very abruptly with my arms crossed over my chest, without resting on anything. This was, I supposed, a way of measuring the muscle strength in my legs. It seemed a silly sort of test, but I was afraid it might be possible to glean significant information from it. I mentally gave myself a fail, maybe just scraped a five.

After this, everything seemed to happen as in a dream. I remember holding a dynamometer in my right hand, which I kept squeezing and squeezing, to little effect. Then I heard the doctor talk about the relationship between fat and muscle (between *my* fat and *my* muscle). I think he said I was slightly dehydrated and that I was mesomorphic.

"So's Arsuaga," he said. "You're both mesomorphs."

I took that to mean we were about average. That we did not, in other words, have very original bodies, which wounded my narcissism a little, and possibly the palaeontologist's, too.

Next he measured my lung capacity, making me inhale air methodically through the mouth, only, when I could no longer take in any more, to blow it back out again in a forceful but sustained way, until I was left more than hollow — empty. I got the sense that I was expelling other things along with the air, ancient things, between good and bad, things of the soul, as there was no way I could be this exhausted from the loss of oxygen alone.

"This," I heard Corbalán say, "is also a way for me to measure the thoracic musculature."

I'd never noticed my thoracic musculature before, but it was unquestionably there, like a breastplate protecting my ribs, which I had also just discovered, having never thought about them before either.

My thoracic musculature, for God's sake. My ribs. My heart. My lungs and their alveoli, the trachea ... So many things! There was no way they could *all* be functioning properly. Any moment now, the doctor was going to discover something disastrously wrong with me.

From time to time, I glanced sidelong at Arsuaga, who, seeing my treatment, I imagined saying to himself: "Forewarned is forearmed ..."

Somebody — I suppose it must have been Corbalán — said between one jotting-down and the next, because he was making a note of everything, "We shift around half a litre of air with every breath we take."

There's nothing worse than being made aware of your body, of confirming that you *are* a body, since right away, along with this information, there comes a whole inventory of limitations and possible pathologies.

I don't know if it was before or after I'd started what was effectively a medical triathlon or pentathlon that the doctor asked if I was on any regular medication. I briefly contemplated lying, as I wasn't keen for the palaeontologist to know all my pharmacological intimacies, but I decided to go for the truth based on the superstitious idea that, if I told the truth, they wouldn't find anything seriously wrong.

"I take one pill a day for my blood pressure," I admitted, "and another for my cholesterol, as well as a sedative to get to sleep. Sometimes two."

He was also interested in the dosages, about which I could at least be unabashed, because fortunately they were very low. Still, speaking aloud those treatments that had already been incorporated into my daily life reminded me once again of the scale of the age I'd reached.

Seventy-five years old.

Wasn't it a bit much, really? Was there any way the engine of a Mercedes or an Audi or Arsuaga's Nissan Juke could have worked this whole time, from 1946, the last century, without stopping for a single moment, as my heart and my lungs had been doing since that date?

"Are you right-handed?" the doctor asked me then.

"Yes," I said.

"Then use your left to pinch your nose, and put this tube in your mouth."

I put the tube in my mouth and I breathed again, emptying myself, and after emptying myself several times further, I got onto the treadmill with my chest totally wired up and with a sort of gas mask on my face, which was connected to another tube, through which I took in the oxygen that was demanded, insatiably, by those two inflatable bags — like shopping bags — that we call "lungs". I imagined the oxygen being carried in my blood to the last cell in my body: thirty-seven billion cells, many of them in my brain, were waiting, eagerly, for that oxygen, without which they would perish or turn stupid. I recalled those lines of Gabriel Celaya's:

> … like the air we require thirteen times a minute,
> to be and in being give a glorifying yes.

Was it actually thirteen times a minute, or was that just poetic licence?

The breathing on a treadmill, anyway, was what they called a "stress test". I'd never done one before. The cardiologist, like the skipper of a boat, was watching my vitals, making a note of the beatings of my heart as it flapped against my ribs like a bird that had just been put in a cage, while he increased the speed or made the treadmill steeper so I had to work even harder. The machine had an emergency STOP button, which I looked at longingly, but being watched by Arsuaga, who was standing behind the doctor, made me keep going and going and going until I couldn't take it anymore and raised my hand for ceasefire just a tenth of a second after the doctor had stopped the machine.

While I took off the mask, panting, and removed the wires attached to my chest, I heard Corbalán saying to Arsuaga, "Ramped Bruce protocol."

151

I assumed that must be what the test was called, named after some Bruce or other who'd invented it.

I'd begun to look at our doctor with the respect with which you look at an airline pilot who, in the middle of some turbulence, comes out of the cockpit and walks up the aisle with an enigmatic expression on his face. Corbalán must have sensed the vibrations of my anxiety because he looked at me and said, "You're doing well for your age."

And he added, in didactic tone:

"The more tired we get, the more CO_2 we produce. We have a routine aerobic metabolism to support our basal activity, which depends fundamentally on fats. While we walk at a leisurely pace, for example, our cells are drawing fatty acids from our body's reserves. However, at the moment when, instead of walking, you start to run, the fats don't supply enough energy intensively and quickly, so you need some other kind of fuel, which is glucose. Glucose supplies more energy, but mainly it supplies it faster. Some requirements are met by fat, and some by glucose. This shift from fat to glucose occurs at the anaerobic threshold, which is the point at which you become breathless."

In my head, I compared the body with a hybrid car, where the engine draws either on the battery or the petrol depending on its energy needs, but I didn't have the nerve to say it out loud.

"Let's see," Corbalán added, scrutinising the charts produced by my stress test: "Sub-maximal test stopped at ninety-five per cent of your heart rate. A very reasonable result. You had no angina pectoris or arrhythmia during the test. No change to the electrocardiogram to suggest any cardiac lesions. Blood-pressure response normal."

I sat down in a chair to get my breath back fully, and

meanwhile I watched the palaeontologist begin his set of exercises. Although I had plenty to deal with myself, I got the sense he wasn't expelling the air very well. Also, when he was on the treadmill, he insisted on running right from the start, instead of walking, and things didn't work out as well as he'd hoped. Then, while Corbalán was taking the wires off his chest, he blamed his poor performance on the mask.

After all this, it was gone midday and the doctor must have had some other appointment, because he told us to come back the following week. "By then," he said, "I'll have prepared your reports and we can discuss them in more detail."

And there we were, one week later, sitting opposite the cardiologist and ex–basketball player, with him on one side of his desk, and Arsuaga and me on the other. The doctor had prepared two folders, one with my name on and the other with the palaeontologist's, containing the reports generated from the tests to which we had been subjected. He had positioned them in such a way that we could follow what they said while he elucidated it verbally. It was rather like he was giving us our end-of-term grades.

"I'm delighted to say you're both in superb shape," he began.

Arsuaga and I exchanged the brief glance of a pair of conscientious students.

Corbalán went on: "In the first thing we tested, your results are the same: the degree of visceral fat — that is, the fat that accumulates and begins to pervade the abdominal organs and that the body calls on last when we're in need of fats. Over the years, this fat takes on certain metabolic activity that encourages general inflammation. It is a bad fat, worse than the usual kind, and it works in a warehouse-like fashion. The ideal for a person at a good weight is for the fat to be consumed as it arrives, the way warehouses should operate in theory: first in should also be first out. If you allow stuff to accumulate, some of the stock gets ruined. In this respect, you're both at thirteen, just above what's considered to be the ideal, which is twelve. Normally, as we get older, we end up over the standard twelve. The two of you are right on the edge, which is very respectable."

Arsuaga and I now looked up from the pages of our reports, which contained countless graphs and percentages, at the face of the doctor, nodding seriously at what we were being told. I think we were a little alarmed that he might have decided to share the good news first, and that the bad was on its way.

"The second thing we look at," the doctor continued, "is the percentage of fat in your body. As I explained, the muscle would be the engine, and fat the fuel. You, Juan José, are 21.3 per cent body fat, and you, Juan Luis, 24.4 per cent. For somebody to have less than ten, they'd have to be a top sportsperson, or at least training very intensively, or alternatively somebody with an eating disorder like anorexia. In the fat-to-height ratio, which is a metric that's used a lot and is much less specific, you're both at around a six, Juan Luis just a touch higher. Another essential piece of data relates to what the muscle mass and fat mass actually weigh. You, Juan José, are at 3.51, and you, Juan Luis, at 2.94. Though nobody's published on this, the inflection point where things could be improved is at around three. As the years go on, it becomes very difficult to improve your muscle. It can increase, but not significantly, because we lose all the anabolic elements: the testosterone, crucially, is reduced, and bones and joints deteriorate. Exercise begins to become less beneficial. So it's vital to keep one's weight under control in later life."

Arsuaga and I exchanged a glance. I was sure he was thinking about the nine kilos I'd passed on to him.

Corbalán looked at our faces and cleared the matter up: "In any case, you're both around three, which is normal. So, from a functional, aerobic point of view, you're both in good shape."

We spent an hour, maybe an hour and a half, going slowly through our reports. Some things I understood, some I didn't. Arsuaga, with his training, understood them all, which allowed him to exchange opinions with the doctor, to which I listened with the fascination of an ignoramus presented with a syntactically well-constructed speech. The palaeontologist, seeking to justify my silences, would occasionally say, pointing at me, "He's a humanities guy."

We did finally find out a lot about our physiology. Analysis of all the parameters allowed us to deduce that our profiles were to be expected for our ages. In some respects, we were towards the lower end, and in others towards the upper, but none were disappointing, though no cause to pop open the champagne either.

Arsuaga was advised he ought to keep a check on his blood pressure.

"Buy yourself a BP monitor," Corbalán suggested. "The best way to do it is in the mornings, after you've been lying down and calm for five minutes."

On that day, Corbalán was wearing a sky-blue shirt under his white coat, and it went well with the decor of his office. I was confirmed in my suspicion that he was actually a movie doctor, the doctor from whom every patient would want to be able to receive their diagnosis, including a bad one. I did engage with what he was saying, but also at moments became disengaged, fascinated as I was by his looks, and also by his bald head, which was of a spherical perfection. His head seemed to be designed for alopecia. And he hadn't made the mistake that so many bald men do, of compensating for the lack of hair with a beard. On the contrary, his perfect shave revealed a well-hydrated skin conveying a cleanliness that seemed, as it were, close to godliness.

"From a health perspective," I heard him say in one of my returns to reality, "our bodies aren't made for really heavy lifting, or for running fifteen kilometres an hour. They're made for finding a balance between what you spend and what you need."

"Palaeolithic man," replied Arsuaga, "would be moving at walking pace or at a jog."

"I think," replied the doctor, "that a Palaeolithic man wouldn't run, because he'd know that running ought to be saved for hunting or for avoiding being hunted. Anyone who leads an active life economises their energy because they don't know if there's a steep hill coming up."

Then came a few more minutes filled with technicalities, followed by another recommendation addressed to the palaeontologist: "If you can slim down a little, you'll get stronger."

"So I did well to lose weight," I butt in.

"Of course you did very well to lose weight," exclaimed Corbalán.

"I'm an active fat man," replied Arsuaga, "and you're a sedentary skinny one."

"No, you're not fat," the doctor said, consolingly, "but you're just about at the limit. If you lost five kilos, it would be reflected in this study. However," he added, "nobody has ever actually defined the ideal. You ask yourself: 'Could I lose ten kilos?' Yes, you can — if you don't eat, if you keep active, allow yourself to feel a bit hungry, et cetera, then you would lose them. But then you might have stopped enjoying life, you wouldn't be having that glass of wine with some friends, or those nice nibbles … That's the point you need to weigh up."

"Right," said Arsuaga with Epicurean shrewdness.

"I tell all my patients," the cardiologist continued, "that I prefer living cowards to dead heroes. When you make plans to lose weight or to exercise, don't set unreasonable targets, because this whole business is riddled with emotional aspects, which is what I also argue with my patients: look, we're here to be happy, not to wander around pissed off our whole lives."

"And to sum up?" I asked, because it was getting close to lunchtime and I could already sense the drop in my water levels, or my sugar levels, or whatever it is that drops at that time, and I was starting to get grumpy.

"To sum up," Corbalán concluded, "the two of you are very close to one another; it's just that in your case, Juan José, with your extra eight years, you're maintaining a particularly high standard. Weak people your age die. Keep active; stillness is death."

I think we left the appointment, each with a report in our hand, a little deflated. Both Arsuaga and I find normality disappointing. Maybe we had a higher opinion of ourselves, of our lungs, of our hearts, of our muscles, even of our fat.

And in the lift, before we made it down to the street, Arsuaga handed me his file, as if it were of no interest to him.

"What do I want with this?" I said.

"In case you want to compare my results with yours, at your

leisure, when you're at home," he said. "For the book."

I took it and we stepped out onto the street, where it was hot and there was an atmosphere of general unease that we might perhaps have been projecting.

"It would have made it more literary," says Arsuaga, "if either your report or mine had been terrible, if they'd told us we were at death's door."

"I don't share your sense of what's literary, but there's a Japanese restaurant over there that's really good. Come on, I'll buy you lunch."

Arsuaga hesitated. He was angry with someone, maybe with Epicurus, I think, maybe himself. Finally, he allowed me to lead the way, although his expression remained unchanged.

"Hey," I said to him while we crossed the road to the shopping centre with the Japanese restaurant inside, "coming to the cardiologist was your idea."

"Do you think it was a bad idea?"

"On the contrary, I like to observe my old age. Well, I don't know if I *like* it exactly, but it does produce a morbid curiosity. One of my favourite books has just that as its title: *A Grief Observed*. It's by a British writer, C.S. Lewis, you might know him. He's writing after the death of his wife, using that method, of careful observation of his grief."

The palaeontologist followed me to the restaurant, which was on the second floor of the shopping centre. It had very few tables, but there was one free. We took it and I ordered for us both: a seaweed salad and two portions of sashimi.

"And a couple of glasses of verdejo," I added.

I like the yellowish colour of verdejo, which might remind some people of urine, but which to me resembles liquid gold. It struck me that my mood improves as Arsuaga's worsens.

I transfer my excess kilos to him, and he passes back his Epicureanism.

Who gets more out of this relationship? I wondered. And I instantly recalled the refrain of a song by Víctor Manuel, whose protagonist asks a similar question:

> Who put in more, both throw it back,
> who put in more, tipping the scales,
> who put in more warmth, tenderness,
> understanding,
> who put in more, who put in more love.

Arsuaga doesn't like Japanese food, I'd noticed this on other occasions. He pretends he does, but then he just pushes things around the plate with his chopsticks to make it look like he's eaten.

"If you like, we can ask for something else," I said.

"No, no, it's fine, thanks."

"Well," I finally pointed out mischievously, "seems like relative to my age I'm doing better than you are relative to yours."

"Because you've lost nine kilos," he exclaimed, "exactly the amount I've put on!"

"But you can't really believe the nine kilos I lost are the same ones you've gained."

The palaeontologist looked at me as if wondering whether I was being serious. He is a scientist.

A scientist.

But scientists have emotions. Then I remembered a day the previous year when we were in Seville, promoting our book. Christmas was approaching and we walked past a lotto stand, where I'd have liked to buy a ticket. I didn't dare, though, for fear of the scientist's scorn; I do know that the likelihood of getting the jackpot is infinitely small, if not zero. The palaeontologist, meanwhile, breezed straight up to the counter and bought a ticket himself.

"I never would have thought a scientist would play the lottery," I said.

"It's a very nice number," he said, showing it to me.

Then he gave me an enigmatic look and put the ticket into his wallet. The next day, I slipped away from the hotel and went to the outlet to buy the same number, which I had retained in my memory with the agility of a young man remembering the phone number of a girl he's just met.

But ours didn't come up.

"The thing is," the palaeontologist said, brandishing his chopsticks, "if we were fleeing from a lion, it would still catch you before it caught me. I'm in a better place than you in terms of functionality. If you aren't convinced, we'll have a running race, or do a few stretches, to clear it up. But if you really need me to say it, I will say it: you're in a better state for your age than I am for mine."

"At last!" I exclaimed.

"I'm not a denialist," he replied. "But the lion would still get you first."

"The fact," I added, "that our bodies correspond more or less to our ages and that we have no chronic illnesses means we've adapted well to life's pre-programming."

"That's according to the theory of ageing and death being pre-programmed," he said. "Which I dispute."

"So old age and death are the result of what, then?" I asked.

"In the eyes of evolutionary biology, they're the result of an accumulation of detrimental mutations, which are expressed at this time of life. The genetic contribution we make to the next generation is pretty minor, not to say non-existent, so whether we live or die has no impact on the continuity of our genes. So, you and I are just in a different stage of physical deterioration; you're further down that track in absolute terms, but not when compared with our respective ages. I'm doing better than you, even if your numbers on the treadmill, in relation to your being seventy-five, may be better than mine."

I honestly wondered whether this eagerness to compete and win was really very Epicurean, but I chose to say nothing.

"What we're trying to work out," Arsuaga explained, putting just the lightest touch of soy sauce onto a piece of salmon, "is what causes ageing and death. Why do they happen?" Then he added, while making as if to bring the piece of fish to his mouth, only to return it to his plate: "If they *were* pre-programmed, like you idealists think, who does that benefit? How's it possible for pre-programmed death to have carried on into our current time, four billion years after the emergence of life? How can natural selection have failed to eliminate it after all these millions of years? How has death, if it does depend on a gene, not been eliminated by natural selection? If it really depended on a gene, any mutation that either deferred it or took it out of the equation would be to our advantage."

"But death," I said, bathing a fillet of butterfish in soy sauce, "has its meaning."

"Please," said Arsuaga, "don't completely soak the fish in soy. That isn't how the Japanese eat it."

"But this is how I like it," I said in my defence.

"Okay," he said, though he wasn't happy. "Those who believe

in pre-programmed death say it doesn't favour the individual but the group — and Pachamama, and blah blah blah … It's of the utmost importance, if we want to put our minds at rest concerning this particular matter, to find out *who* benefits."

"In that case," I said, "dying would be a form of altruism."

"Welcome to the world of warm fuzzy feelings, but that can't be it, it just doesn't fit. It's incompatible with evolutionary theory. It'd be like you asking me if somebody could be allowed a pass when it came to the laws of gravity. Sorry, but no. For selection to have brought death about, somebody would have to benefit from the dying, because things that are detrimental never get selected. But, absolutely, there's a contradiction in the fact that the maturation process is pre-programmed, while death isn't. Yet death benefits no one."

"Except the community," I suggested timidly, while gesturing to the waiter to pour me another glass of verdejo.

"Well, those are exactly the romantic, Kropotkian theories we're trying to get away from. Don't get all 'Channel 2 documentary' on me — you know, with that voiceover they do, explaining everything away in a flash."

"Okay," I conceded obligingly (the verdejo had gone to my head a bit), "everything's the fault of the mutational burden. I don't really know what the hell that means, but I do love the syntagm 'mutational burden'. It could be the title of a novel, or even better, of a poem."

"It's what Medawar says," said Arsuaga, refusing a second glass of wine. "That the mutational burden is responsible for producing decrepitude."

"Here's to the mutational burden," I raised my glass.

"When it comes to the production of sperm and to ovulation," Arsuaga went on, "mutations occur. We've already touched on this. Let's imagine one were to happen that meant — that is, it pre-programmed a mutation — that you're going to have type-one diabetes before you reach adulthood, before you're able to reproduce. It develops when you're twelve, for example, and you die without having produced any offspring. Natural selection has wiped it out. Now, instead, imagine the effects of this mutation coming about when you're eighty. Natural selection can't then get at it, can't eliminate it, because it never sees it. Remember that, in nature, those of our species never get to be seventy years old, or only very few do, because the individual is fragile and the members of a generation drop like flies. There's so few of us left now, they'd say."

"I know decrepitude is an invention of culture and that in nature you're either in your prime or you're dead, but please do let me — aged seventy-five — enjoy this incomparable raw tuna and this golden verdejo."

"People," Arsuaga went on as if talking to himself, "need to find a meaning for their lives, and their deaths. The only way to give life a meaning is by inserting it into a cosmic plan, such that people have a reason to be alive, a plan in which everybody plays a role, however modest. A three-year-old dies — such injustice, no? But along come the Christians to tell us that this death is all part of a plan that's beyond our capacity to understand. God writes in straight lines, but the book he writes in has crooked lines. Don't try to understand it, just accept it. It's like trying to make a dog understand Kant ... Which is the worst sin of them all?"

"Pride," I said.

"In comparison," said Arsuaga, "all the others are as nothing. It isn't good to go and steal things, but you'll be forgiven. Adultery, that isn't right either, but go and confess and you can draw a line under it. Avarice, anger … all will be forgiven. But pride is unacceptable because it has this link with knowledge, with the desire to know, the arrogance of doubting that there's a cosmic plan."

"Can't there be a cosmic plan for people who believe in evolution?" I asked.

"Do you think such a plan included the fact your parents would inevitably meet and give birth to you?"

"Rationally speaking, no, I don't, but it consoles me to think so. Don't you want another glass?"

"No, thanks."

"Well I do, but you're also letting me have all the food …"

"Are you one of those people who thinks nothing happens by chance?" he pressed me.

"It's just that rationality and feelings are often kept completely separate, so you can believe or not depending on which side of the bed you get out on. Today I woke up an optimist."

"Remember what Oscar Wilde said, though: the basis of optimism is sheer terror. Curiously, it's actually harder to be a materialist at this point in the twenty-first century than it was at the end of the nineteenth century, or at the beginning of the twentieth. Everyone seems to believe in 'energy', all that jazz, or even the horoscope or tarot. Me, I'm just a party pooper, a real prophet of doom — the gloomy bastard going round insisting there's nothing after all this. *That* guy. And nobody wants to hear it, because people need to believe in something."

"I'm very interested in what you say," I point out, genuinely moved by Arsuaga's fearlessness, which is the fearlessness of a sober poet, given that being sober and clear-eyed is the worst of combinations.

"Materialism is verboten, Millás. The worst thing ever, people think. We have to believe in the Amazon, in saving the Iberian lynx, in the tropical jungle, in whatever — it's a universal cry. Please, somebody save us from materialism!"

"But I come from a materialist intellectual tradition!"

"Well, then you've clearly forgotten it. When the Guadarrama National Park was created, the worst people involved weren't the people trying to sell the idea, because them you could negotiate with. But you couldn't negotiate with the die-hard ecologists, not ever. Why?"

"Because they had faith."

"They gave absolutely no quarter, it was all 'appeasement', 'betrayal'. Ah, such purity ... Whereas I — who played a tiny part in the park's creation — got accused of having had the developers build some villa for me, goodness knows where."

"That's what you get for sticking your oar in. It's like my mother used to say: don't stand out, blend in."

"Everyone had to be consulted, the farmers, the hoteliers, the developers ... We had to get our hands dirty, while the ultra-greens, seeing as they didn't negotiate, stayed whiter than white. If you don't believe in a religion, choose a cause."

"Or a nation," I suggested.

"It'll do you no good, coming over to my side," he said. "Last summer, on the beach, weighing this project up, thinking whether we should do something together on old age and death, I thought I ought to give you the chance to save yourself."

"But I don't want to save myself. I want vegetable tempura and another verdejo."

"'My goodness,' I thought. 'I'm really going to fuck up Millás's life. He still believes there might be a meaning to it all.'"

"I've got days for meaning and days for meaninglessness. On the meaningless days, I phone you."

"Well, I want that much to be clear, at least: I did give you the chance to save yourself."

"It's perfectly clear. How about another glass now? This verdejo tastes of herbs and nuts, but also a bit of honey, too."

"Alright, now we've made it clear that if you don't save yourself, you've only got yourself to blame — sure."

NINE

Food for the lion

At Arsuaga's gym, he's allowed a number of guests each year, so one day he called me. He said, "Has your tracksuit been washed?"

"Not yet," I replied, "I've only worn it three times."

"Well, get it ready for Tuesday, I'm going to take you to a gym in my neighbourhood so you can do a bit of cardio work. You need to combine your diet with physical exercise, and that's something that needs to be learned."

"The 'whole niche'," I deduced.

"You got it," said the palaeontologist.

The gym was near his house, so we met outside his front door first thing in the morning and strolled over. There was equipment here and there, and a considerable number of glorious-looking bodies building up their muscles, or toning them, or whatever, in sportswear of the most suggestive kind. I noticed an older man unhurriedly pedalling an exercise bike as he read the day's edition of *ABC*.

We went on the cross trainer. Two sessions for me, of twelve and thirteen minutes respectively. According to the information I was given by the machine's computer, I managed to burn off seventy calories working at three successive difficulty levels over almost a kilometre. I told Arsuaga that I could push myself a bit more, and he programmed the machine again, and I climbed back on, ready for anything. Soon, seeing me getting out of breath, he asked if I wanted to leave it there.

"I don't want you collapsing on me," he said, "and to have to drive you to the emergency room. That's enough for today. Let me take you for breakfast."

The palaeontologist, who was seeing me getting thinner by the day, seemed to think I was malnourished, and maybe he wasn't wrong.

We went to a nearby cafe, where we occupied one of the outside tables; though it felt like it was going to be a hot day, it was still cool enough at that time. Arsuaga ordered churros and was brought three gigantic ones, and I had toast with York ham.

"Concepts," said the palaeontologist.

"What do you mean, 'concepts'?" I asked.

"The concepts of leggings and neoprene," he clarified. "Us white folk, we look better, more attractive, in a pair of black leggings."

"How so?"

"Okay. I wanted to talk to you about the skin, which in turn leads us to the matter of sweating. You've had a decent sweat on the cross trainer."

"I really have," I said, "but I'm taking a liking to it. It's a kind of asceticism."

"Right, so a human has the same number of hair follicles as a chimpanzee."

"You wouldn't think so."

"But let's make a distinction between the fine thing that's

more like *down*, and the *hair* on our heads, armpits, and faces … A linguistic nuance that, though I don't know if it's totally correct, is still helpful. In any case, we have the same number of follicles as any primate, chimpanzees included. We aren't naked monkeys, we're downy monkeys."

"Why is it some people have hair on their backs and some don't?" I suddenly wondered, remembering a particularly hairy uncle.

"We all have follicles, which is the structure that the hair comes out through. Some people produce long, thick hairs, and others nothing but a light down, but we all have the same number of hairs. Gym leggings replace the tough kind of hair we're supposed to have, that's why they feel so good. That's my take on it, anyway. If the down on our body were hair instead, we'd have leggings made of hair. In short, we've replaced hair with down, and that's why we sweat. Sweat is our way of regulating our body temperature."

"And chimps don't sweat?"

"Far less. We've got ten times as many eccrine sweat glands as a chimp."

"Eccrine?"

"That's what they're called. They're tubes that bring the sweat to the surface, in order to keep us cool. When the sweat evaporates, the body cools down. But there's a cost to this. If you went out and spent a day doing a lot of exercise in the baking African sun, you could lose between ten and twelve litres of sweat, of water. That's a hell of a lot. So we can't do that without drinking, because if you lose that much liquid, you'll die."

"You need a spring or a river," I deduced.

"Or a water canteen. We depend on water because our

cooling mechanism is based on sweat."

"The water ends up being like the liquid in air conditioners," I suggested.

"Now for the science," the palaeontologist went on. "Sweat, when it evaporates — that is, when it goes from a liquid to a gaseous state — absorbs energy. It's the same as when you boil an egg: in order for it to change state, you put it over a flame, right?"

"Right."

"When it evaporates, the water sucks up this energy, which is provided by the flame. Heat, and being physically active, makes the water inside our bodies evaporate, but the water absorbs energy from the skin, thereby cooling it. Humans are virtually unique in this respect. Other animals cool themselves down by panting."

"Which you see very clearly," I said, "in the case of dogs."

"Horses are the one exception. They sweat through their skin, but we're the species that sweats the most — by a long way. This is very significant, and here's why: we aren't the fastest of all creatures, but we do have the best stamina, and it's all because of our cooling system. When hunting, as long as we have access to water, we can wear out our prey. Which is to say, we have need of a prosthesis called a water canteen."

"How can we tell when we're dehydrated?"

"One very common symptom is disorientation, because the brain only functions within certain very restricted limits in temperature. When we've got hypothermia, it doesn't function, and if we go above thirty-nine degrees, we'll become delirious. Many of the people found dead in the Australian outback are found a long way from any paths. The lack of water disorients them, and they become lost."

"So it's a good system," I concluded, "so long as you recharge it."

"Indeed."

"And when were canteens invented?"

"This is where we move into the realm of speculation. Were there canteens in prehistory? Did they have a perfect knowledge of where all the rivers were, the ponds, the springs? Whatever the case, dependence on water was still total."

"And in fact," I pointed out, "cities were built around rivers."

"A million, or a million and a half years ago, were there canteens? Maybe. The Bushmen always carry some with them, or they leave them buried in strategic locations."

"The Bushmen," I repeated like an echo, sipping my tea, as the conversation had started to make me thirsty.

"When we exercise," continued Arsuaga, "body temperature rises, as does heart rate, because the two are connected. If you want to know whether you've got a fever, you check your forehead or your pulse. My grandfather, when we were little, used to take our pulse and say: 'thirty-seven point five degrees'. There's a direct link. You've done some cardio work today, and therefore your heart rate and your pulse have gone up. And your body's reacted, cooling itself by sweating, a marvellous mechanism, one of evolution's great inventions."

"It really has," I agreed.

"As is always the way when it comes to evolution, this cooling system is bound up with a thousand other things. For example, standing up favours the cooling process because, when you do that, you increase your body surface and your whole body comes into contact with the air. When we went from being quadrupeds to bipeds, we moved away from the ground, which meant we heated up less, because the ground's very warm. Some claim the whole reason for our moving to two feet was for thermoregulation. I see them rather as synergies. I don't believe we moved to two feet *in order to* thermoregulate, but there's no question that the upright posture does favour thermoregulation: you get the benefit of the breeze, when there is one. The whole of a quadruped's back is exposed to ultraviolet radiation when the sun is at its highest, while in our case, only our heads are exposed, and we've got our hair to protect them. The bipedal posture and the temperature-regulation mechanism make us superior to all the other animals in the middle of the day. Some people point to this as an ecological niche that we've known how to exploit."

"In the middle of the day," I pointed out, "lions in nature documentaries are always shown lying around like they're in some eternal siesta."

"At that time of the day, not even the insects move. In Africa, all the animals are either crepuscular or nocturnal. They come out either at dawn, twilight, or night."

"And how are the churros?" I asked.

"Fantastic. Want a bit?"

"No, thanks."

"Also, just to say," said Arsuaga, "you shouldn't wear that T-shirt to the gym."

I was wearing a T-shirt he'd given me, with a Neanderthal design I really liked.

"What's wrong with my T-shirt?" I asked.

"It's made of cotton, and the sweat will get trapped in the fabric. Imagine you climb a mountain and when you get to the top you're covered in sweat. You stand still for a minute, the sweat cools, you catch pneumonia and die. Bear it in mind, if you're really thinking about taking up exercise to go with your diet. You have to wear a kind of fabric that expels the sweat, like my shirt here, which, if you look closely, has got all these pores, these tiny little holes, plus it's figure-hugging. It's like a second skin. You even get these for going up Everest. When you're climbing mountains, you have to wear something figure-hugging, because it maintains body temperature and means you don't get soaking wet."

"Maybe one day we can go to the Decathlon store together and you can show me around."

"No problem. They aren't completely perfect: they let the liquid of your sweat through, but trap the salts, which means they can get a bit smelly. Although it isn't actually the salts that smell, but rather the bacteria, which start to decompose."

Whenever the palaeontologist deploys such apparently unimportant knowledge, it makes me want to get up and clap. I love precision. Poetry is precision, hence the lyrical virtues of scientific discourse.

"So the smell is because of the bacteria that are left behind on the fabric," I said, as a nudge for him to keep going.

"But leaving that aside … for every ten to twelve litres of liquid we lose, we also lose three teaspoons of salt. In other words, you have to hydrate and take on salts, hence the invention of isotonic drinks. You and I, we aren't running marathons, so we don't need them. But really-high-level sportspeople do."

"Cyclists," I said, since they're what occur to me.

"And there's something else I need to tell you about the skin: do you know what colour a chimpanzee's skin is?"

"White?" I asked.

"Very good. White like a piece of parchment that's never been exposed to ultraviolet radiation. In human evolution, when we lost our fur, our skin turned dark; when we exposed ourselves to that radiation, we had to develop pigment. A chimp has barely any melanin. The same goes for European people. But our African predecessors did, and present-day Africans do. Otherwise they'd get lots of skin cancers."

"So the chimpanzee's skin is white, and ours is dark."

"Yes, that's well put, because the normal thing for people is to be dark. Being white is the abnormal thing. In northern latitudes, melanin is virtually an inconvenience, because in those places there are barely any problems with skin cancer, but instead we have problems with vitamin D, which is synthesised by the skin."

"I take one vitamin-D pill a month," I remembered. "Every time I do a test, it tells me I'm deficient."

"Because of precisely what I'm saying."

"And why does the skin age?"

"Because it loses elasticity, collagen."

"Does collagen not regenerate?"

"No, we don't have genes that do that, because by the age you stop producing it, you ought to be dead. Alzheimer's, along with other illnesses people suffer in old age, comes about at a time when we ought to be dead. That's why natural selection has failed to eliminate them — we've talked about this before."

"Don't assume it's so easy to take it all on board," I said. "Especially when you ought to be dead."

"Now," said Arsuaga, "I'm going to ask you a question: if a lion had come into the gym when we were there, who do you think it would've eaten first?"

"Well, me," I admitted. "I was older and therefore slower."

"Exactly. Who next?"

"The man reading the paper on the exercise bike."

"Out of us two."

"You."

"Okay. I'll tell you a joke that often gets told in university. A pair of African hunters are out one day and they suddenly find themselves face to face with a lion. They glance at their rifles, but they're all out of bullets. They start to panic, they don't know what to do. One of them crouches down and starts doing up his trainers. The other one says: 'What the hell are you doing?' 'My laces.' 'You don't think you can outrun a lion, do you?' 'No, but I think I can outrun you.'"

Instinctively, I looked down at my shoelaces. Arsuaga smiled and added, "That's natural selection in a nutshell: be faster than the other guy. You'd go down first, being the slowest, and the next day the lion would come back and eat me. The probability of being eaten increases the older you get — that is, with a loss of function."

"Right."

"Now," he continued, after a sip of his coffee, "I feel like it's important to establish a distinction between ageing and old age. I'm ageing. You're old."

"I feel like we've talked about this before," I said.

"And what does getting old consist of?"

"It consists of it becoming more likely that the lion will eat you," I replied.

"Exactly. I'm not old, but I run less than I did a decade ago. Over the course of our lives, we lose certain faculties. People break Olympic records between the ages of twenty and thirty; it's far less common after that. One way of explaining the ageing process is through athletic achievement."

"However," it occurred to me, absurdly, "the average age for popes in Rome is seventy-six. I could still be the Pope."

"Sure," the palaeontologist conceded with a pitying look.

"There are lots of different ways of expressing the gradual loss of faculties. I'm old compared to when I was twenty. But, I still say, it's important to distinguish between ageing and old age."

"Old age is a state," I concluded. "Ageing is a process."

"Just so we're clear," the palaeontologist went on, "old age is the place we find human beings like you, who in nature would be dead. Remember that maxim we've talked about, and that we'll certainly come back to: in nature, you're either in your prime or you're dead. The test tubes in Peter Medawar's example are just as breakable to begin with as they are at the end. They break by accident, until there are none left. And it's after there are none left that those genes that natural selection couldn't eliminate are expressd. Natural selection doesn't get sight of them, because there's nobody around in whom these genes are expressed."

"It means that I, as someone who's managed a five-year victory over death — because I'm seventy-five, when the longevity of human beings is around seventy — can start to be afflicted, from one moment to the next, by illnesses that don't come up on natural selection's radar."

"Today, I think you've got it. So why has natural selection not eliminated the cataracts suffered by an eighty-year-old gentleman?"

"Because he should have been dead ten years ago. A lens in good condition lasts just as long as it ought to, not longer. It would be wastefulness, a pointless expense, as Ford would say, the one with the cars."

"Very good. That eighty-year-old has escaped the laws of natural selection, but is now exposed to cataracts, to Alzheimer's, and to a whole long etcetera of illnesses well known to the inhabitants of residential homes. This is old age,

Millás, the genes. You've got that, right? I've noticed that you sometimes just pretend to have understood ..."

"It's true, I do sometimes pretend, because I'm ashamed of not understanding, but on this occasion I think I do actually follow."

"Well, good. Now we've got our heads around this, I need to explain a certain refinement to Medawar's test-tube theory. It's an interesting one — the antagonistic pleiotropy hypothesis."

"I'd better ask for another tea. Will you have another coffee?"

"Sure. Write it down: antagonistic pleiotropy. So, what we've just been talking about — there's a set of detrimental mutations that we carry in our genome and that are only expressed in later life — we'll call that the *mutational burden* theory, since you like the words so much."

"The time of life," I said, just to be sure, "when natural selection isn't in play."

"Right, because, by then, the group of hunter-gatherers has been decimated. Hunger, cold, thirst, lions, bears, wolves, falling from great heights, thunderbolts, other humans ... Gradually, each and every member of a generation goes down. But we're going to refine what we've said. If I were a test tube, the probability of me being eaten by a lion would be the same as the probability of a twenty-year-old being eaten by a lion."

"Yes, because, according to Medawar, you can only be either perfect or broken."

"But test tubes don't get old, they don't lose their faculties, whereas I do. We saw this in the museum. A single gene produces various effects: and we call this capacity 'pleiotropy'. A single gene does various different things, and often they're antagonistic. The theory of antagonistic pleiotropy says there are genes that, when we're still young, benefit us, but when we get older, do the opposite. George C. Williams, who we've already touched on, tells us to picture a gene that produces calcitonin, a hormone produced by your bones to lower your blood calcium. This same gene then leads the arteries to become calcified — arteriosclerosis — when you get old."

"Hence the antagonism," and I made a note.

"It's the price you pay for the exuberant fertility of youth."

"As if the same gene," I suggested, "hosted both a life impulse and a death impulse. Eros and Thanatos, the two basic impulses that, according to Freud, are found in human beings."

"Natural selection has two opposing forces," continued Arsuaga, "pulling us in opposite directions: one telling you to have all the children you can right at the start, because you don't know if you're going to get a second shot, and another telling you that you shouldn't die immediately thereafter, in case that second shot does indeed come along."

"There's got to be a balance."

"And natural selection attains this balance. As I've already told you, in the case of Pacific salmon, for example, given that their mortality rates are very high, and their chances of spawning a second time around are very low — maybe because they have to go up rivers that are both particularly long and fast-flowing — the first of these two forces has won out. The force that tells you to give your all the first time around. Whereas with Atlantic salmon, which are able to swim upstream more than once, the other force dictates that they hold something back, so as to have more opportunities in the future. There are two forces, anyway, two selection pressures. One: the more breeding seasons you have, the better it is for perpetuating your genes. The other: just in case, give pretty much everything you have at the first opportunity. One or other is stronger in each species, and corresponds to their chances of dying, which are also the chances of living. So, if mortality rates are very high in your species, go all-out right at the start, like octopuses do. And if they aren't so high, pace yourself, so you get various shots. We humans, though our reproductive capacity does diminish over the years, we're still able to have children when we get older. But the same genes that in their day enable us to have lots and lots of children can lead to cancers in later life. Think about the prostate! The same things that engendered life later become the causes of death. Natural selection doesn't deal in miracles. From the age of fifty onwards, we start to pay the price for this reproductive capacity with chronic illness, and from seventy onwards, if there's no special care available to you, you're dead."

"Gosh," I said, as I attacked my second cup of tea.

"I'm doing fine," said Arsuaga, "although I suppose my sperm count, not so much (and let's not joke about erections). Plus, I've got life experience, and that makes up for the loss of certain faculties. What I mean by that is I don't go around doing stupid things, which is often a reason young people die. I can still run a bit. Overall, with one thing and another, I've still got two or three more years of being able to get away from the lion. In nature, give it four more years, and I'd be dead."

"That's why we've uninvented nature," I said, "and we've given birth to culture."

"But at the cost of having to go through decrepitude," replied Arsuaga. "Decrepitude is the bill you get left by the genes that previously made you dark and handsome, or made you a stunning blonde, with lots of children to show for it. You were the strongest warrior in the tribe, you say to yourself, contemplating the youngsters. But it's catching up with you now. That lion's got his eye on you."

"Just so we're clear," I concluded, "there's one way of looking at it, which is that of Medawar's test tubes, according to which, if there were no such thing as death by natural causes, we'd die because sooner or later a roof tile would fall on our heads; and then there's a qualification, introduced by George Williams …"

"And both are correct," pronounced the palaeontologist, stirring his coffee. "Williams's qualification explains the loss of faculties that comes about over time, after the age of thirty, and into later life …"

Then, gesturing for the bill, he concluded our conversation: "And, Millás, don't forget: lions never miss a meal."

TEN

Slowing the pace

I was lying on my back, on the couch, hands on my chest, telling my psychoanalyst (or myself) about the curious relationship between genes and the soul as regarding the contradictory life and death impulses, when I felt my phone vibrating in my pocket. I knew it was Arsuaga because it often happens that I get called by the person I'm talking about at that moment.

"I'm getting a call from the palaeontologist to whom life doesn't owe anything," I said, pointing to my pocket where the vibration was coming from.

"How do you know it's him?" asked my therapist.

I preferred not to tell her that these coincidences do happen to me, because I know that she, though she doesn't say as much, thinks I'm easy prey to magical thinking.

"Just instinct," I replied.

"And what it is you call just instinct?" she insisted.

I changed the subject because I realised she wanted us to talk precisely about this problem of mine with magic; a problem, incidentally, that makes up one variant of persecution complexes. I won't deny that I'm a bit paranoid — but only if people aren't hounding me for such an admission. In other words, I believe in synchronicities, a Jungian concept that my psychoanalyst, I thought, might find hurtful, seeing as she's a Freudian through and through.

Par délicatesse j'ai perdu ma vie.

When the session ended, back out on the street, I confirmed that I did indeed have a missed call from the palaeontologist. I dialled, and he picked up at the fourth ring. Of all the people I know, he is the one who takes longest to answer his phone, as if he doesn't care who's calling.

"I knew it was you even without looking," I said.

"Did you? How?"

"Just instinct," I replied.

"And what, to your mind, is instinct?" he replied in turn, because he, like my psychoanalyst, hates any manifestation of magical thinking.

"The same thing that makes you buy a lottery ticket at Christmas."

"But instinct doesn't work for me, because it's not my thing. I buy a lottery ticket precisely to stop it from becoming my thing; that way, I don't succumb to the temptation of complete delusions," he said dryly.

"Whereas for me, on the other hand, it always works," I maintained. "The phone rings, I tell myself it'll be So-and-so, and ninety-nine per cent of the time it's So-and-so."

"Okay," he said, considering the question settled. "Congratulations."

"Thank you."

"I was ringing because there's something you won't believe."

"I believe everything. Try me."

"Remember what we were saying about cells and cellular ageing?"

"Perfectly."

"Well, it turns out I know somebody who does a kind of blood test that's unique in the world, which can be used to tell the difference between your chronological and your biological age."

I hesitated, then said, "And who is this person?"

"You needn't worry: she's a professor at Madrid Complutense."

I still wasn't certain.

"Did you say she's a professor?" I asked, just to be sure this wasn't something too esoteric, a ridiculous worry when Arsuaga was the go-between.

"A professor, yes. Why?"

"It's just that this whole thing about blood tests that are unique in the world sounds strange. She's not a bloodsucker, is she?"

"Millás, please. This is Mónica de la Fuente we're talking about, a highly respected researcher who's spent more than thirty years studying the subject, which is hugely relevant for our book. She can tell us, using these highly complex blood tests, the real age of our bodies. And although she's extremely busy, I've got her to fit us in for an hour on the 29th."

"I don't know if I want to know the real age of my cells," I said.

"That means you also don't want us to write the book," he said, menacingly.

"This book," I said, "has forced me to see things in myself that I'm not at all happy about."

"The truth may not be the sweetest pill …"

"Remember what happened to us with Corbalán: it turned out you're older than you should be at your age."

"That can be fixed. I'm working on it. I've lost two kilos already."

"And what if she finds something bad?" I asked.

"She's only going to tell us whether our chronological age coincides with our biological age. It's a unique opportunity, honestly."

Doctor Mónica de la Fuente is indeed a professor of physiology at the Complutense University of Madrid (I'd checked everything out online because I don't like just any old person taking my blood). I figure she is Arsuaga's age (about sixty-six or sixty-seven), though she looks younger and conveys huge amounts of energy and efficiency. From the off, there is something about her I can't help but like, and she always expresses herself with vocabulary that's precise and syntax that's nicely rhythmical. Being close to her makes me feel safe.

We've met very first thing in the morning, at the Faculty of Medicine, the hallways of which are empty owing to the time of year (29 June). Before getting the jab, we sit around a table at which she explains that the blood test to which we are going to subject ourselves is primarily used to see how our immune system is working.

"How your cells are," she adds, "that allow you to defend yourselves against infections, cancers, et cetera."

I don't like the fact she's started talking about infections and cancers, but I keep listening.

"After many years' research," she continues, "we've seen that the functioning of these cells is the best indicator of health. Other researchers had already detected it, but we were able to provide confirmation. We have also verified and confirmed that some parameters obtained from these blood tests make it possible to see how quickly the individual is ageing. They can determine, in short, a person's biological age, which often doesn't coincide with their chronological age."

Arsuaga and I exchange a look. I give a gesture indicating some uncertainty, possibly a gesture of "let's get out of here". I don't want to know how quickly I'm ageing or what my biological age is. Besides, I'm afraid that the blood test might reveal other things that are even worse. But Arsuaga seems keen, and he gives my knee an encouraging little tap under the table.

"The only thing we do here," the doctor continues, when she notices my hesitation, "is draw some blood, take it to the lab, and through a rather complicated system, separate the immune cells, the phagocytes, the lymphocytes, and the ones we call the *natural killer* cells. They've got that nickname because they're what our organism uses essentially to destroy cancers. Although they increase in number with ageing, their ability to reach and destroy tumour cells is reduced."

Natural Killer sounds like the name of a Tarantino movie, and I've got a bit of a prejudice against him. On the other hand, I don't like how often the word *cancer* is cropping up. I cross my fingers whenever I hear it, but I've got to keep my hands hidden so no one will notice.

"So you measure their capacity to destroy tumour cells," says Arsuaga.

"We analyse this activity in the natural killers," explains de la Fuente, "but also a whole series of functions in each of the other cell types, through a battery of tests in the lab. We've been doing it for more than thirty years and we've been able to confirm that someone with the best results in these functions is healthier, has fewer illnesses, and lives longer. The reliability of these markers for determining biological age is almost ninety per cent."

"Is it possible," asks Arsuaga, "for someone who's seventy years old chronologically to have a biological age of eighty, for example?"

"Or fifty," replies the doctor. "We've done hundreds of tests and, to our surprise, the most common thing we find is healthy people of thirty or forty whose biological age is really sixty or seventy. We started with a large database on which we evaluated how a number of functions were doing in people of different ages, both men and women, because there's a bit of variation depending on sex. Then we extracted a mathematical model from the database, which shows the performance levels of those functions that are most key to each individual's condition. They don't all have the same value. Finally, we were left with the five parameters that make up the *immunity clock*, which determines biological age. That is, how fast you are ageing. And we've validated this in many ways."

"With mice?" I wonder, because I'm starting to feel like a lab animal.

"It's not possible to draw enough blood from mice to analyse these functions, let alone to do it without killing them," says de la Fuente. "You can do it in the spleen and the thymus, which is where there are a lot of immune cells. But that's no use to us, because we want to go on evaluating each animal until its natural death. So then we confirmed that the peritoneum of little mice is where all the immune cells are, the same ones that you find in the blood, the spleen, the thymus, and so on. And they are equivalent to the ones you get in peripheral human blood. These cells from the peritoneum can be obtained easily from each animal, without even needing to anaesthetise them.

We've had a lot of experience carrying out that procedure. By injecting a tempered saline solution into the peritoneum and doing a little massaging, we get the whole suspension of immune cells. We analysed the functions of these cells as the animal ages, and we saw that the trajectory followed by their functions is similar to that in humans. The difference is that our life expectancy is much longer than that of a mouse, which lives two or three years. We've published this study already."

"And the mice don't suffer?" I ask, completely identifying with them.

"It's a very simple process, not too invasive. Like when blood is taken from us."

I repeat to myself some of the terms I've just heard, the ones that particularly resonate (it's a defensive ritual I use in conversations that scare me). This is a part of the litany: "little mouse, spleen, thymus, peritoneum, anaesthetise, massage, tempered saline solution ..."

Once I've completed my ritual, I get hooked on the doctor's speech again.

"Since the mice only live to about three, their ageing process begins at six or seven months, and it's quite possible to wait for each animal to die. Thus we have also been able to produce mathematical models with the mice. These have been published under the question: 'When will my mouse die?' We analysed the functions of the immune cells of the peritoneum when the animals are adults, we applied the model and predicted when each one would die, and it turns out to be right on the money."

Oh God! I say to myself. How long will my mouse live, and how long will Arsuaga live, and how long will I?

Turns out, it's right on the money.

A bad business.

We shouldn't have come.

"So," asks Arsuaga in a technical tone, or maybe it is an Epicurean tone, I'm not sure, "will this test tell us when we're going to die?"

"Yes, more or less," she says with a wicked smile. "But this can vary positively or negatively if you alter certain lifestyle habits. That's the advantage of finding out your biological age, that you can influence it."

You can influence it, I think, if you repent. If you repent your wine, your saturated fats, your sedentary life, your cheap burgers, your freezer croquettes, your processed chicken, your bacon, your coffee, your adolescence, and your youth. Inwardly, I ask forgiveness of biology — my biology — for all of this, and swear to mend my ways if it's not too late, at my seventy-five years of age, to be forgiven.

Arsuaga continues with his technical questioning, revealing no emotion whatsoever (the advantage, I tell myself, of having read Epicurus): "What's the age in human beings when these variables are at maximum functionality in the immune system?"

"If you're a woman, full potential happens around forty. If you're a man, around thirty."

At that point, I escape from my obsessing for a moment and I hear myself say, "I haven't really understood what it is that happens to women around forty."

"Arsuaga has just asked me," explains de la Fuente patiently, "when it is that immunity is at its strongest, when one's defences are at their best. I was saying that for women it's around forty, and in men, around thirty. All under normal conditions, of course. Imagine a person who's had a bad adolescence, who not only has a terrible diet but has severely overeaten."

"Done."

"What's done?" says the doctor.

"I mean, I've imagined that adolescent, I've got him in my head."

De la Fuente looks at the palaeontologist as if asking him *who is this guy you've brought me* and then continues: "This adolescent will not arrive at the beginning of old age in a standard condition. He will reach adulthood already somewhat 'aged', so that if he doesn't do anything to remedy it, his ageing will be faster and he will die earlier than he really ought to. We've studied this and published papers on it."

"With this test you do, are you going to tell us what age we're going to die?" I ask again, because of my nerves.

"We're going to tell you how quickly you're ageing," says the doctor, all sweetness and light, "irrespective of your *chronological* age. If you're older biologically than you are chronologically and you do nothing to rectify that …"

"If you stop smoking," I offer as an example, as I do have a cigarette from time to time, always in secret, even in secret from myself (though officially I don't smoke).

"That's a good example," says the doctor. "We have on occasion repeated the test with a one-year gap, and the subject went from having a biological age of seventy-something to thirty-something. A lot of celebrities learn their biological age, and it annoys some of them. I remember one of them saying, 'I don't get it, my triglycerides and cholesterol are good, I really take care of myself.' And I told him: 'That's not all there is to it. In order to be well and to age more slowly, you don't just need to think about your diet; it's about attitude, too. Sadness can have an impact. In your case, that's what you ought to change: when you showed up here, you were very downcast.'"

Diet, I remember, is a part of the niche. What I'd never imagined is that state of mind is a part of it, too.

"Of all the factors we've seen here," the doctor concludes, "I'd say the emotional is the most important."

"The emotional!" I exclaim. "Your blood tests are able to measure emotional state?"

"Totally," she says.

"I don't believe it," I say.

"Well, you will. I've spent thirty-five years teaching a course called 'Psycho-neuro-immuno-endocrinology'. I know it sounds like a tongue twister, but it's a science that studies how our emotions, or our thoughts — which end up generating our emotions — influence the immune system. If we're sad or stressed, for example, molecules get produced in our brain that reach the immune cells, and the immune cells start to work less well, they don't defend us properly anymore. While when we're happy, the opposite happens."

"Psychosomatic," I say.

"Of course!" she says. "The mind isn't separate from the body! Our emotions, much more than what we eat or the physical activity we do, will control our immunity and thus influence our health."

"So what do we have to do," I ask, "if we want to stay in shape?"

"Well, personally, I'm pretty bad where diet's concerned, because I love eating and I think I eat more than I should. I haven't got time to exercise, though I do try to walk everywhere and I'm very active, but what I try most of all is not to let my mood go, not to decline. It's not always easy, but I make every effort because I know that's what will do the most to ensure I don't get old so fast."

"And those cells, the natural killer ones, are they in the blood?" I ask now.

"Yes, but they're in lots of other places, too. The advantage of the immune-system cells is that you don't have them only in the thymus, in the spleen, or in the ganglia, but they recirculate, so that whenever you draw blood you can always find them. They're out on patrol."

"And these cells," says Arsuaga, somewhere between a statement and a question, "detect a tumour in the liver and ride the bloodstream to the affected organ, and the blood vessels release them there so they can act."

"Of course!" replies de la Fuente. "The immune cells that move around our circulatory system, whenever they detect a pathogen or some other problem in a tissue, they pass between the cells of the blood-vessel walls and start to act wherever they're needed."

"So it's possible," I say, "that we've had cancers without even knowing it?"

"Constantly," replies the doctor. "Our cells are always turning malignant, but they don't turn into cancer because that's what our immune systems are for: eliminating them. It's only if there's some lapse in this vigilance that a cancer appears."

"And this blood test you're going to do on us," I continue, "it's unusual, isn't it? Arsuaga told me you can't get it done just anywhere."

"Correct," replies de la Fuente, "and it's very important because it allows you to act pre-emptively, to change some aspect of your lifestyle. No illness affects one hundred per cent of the population. Ageing does. But that's not an illness, it's a physiological process, it affects us all. And it's good to consider it: seeing as it's inexorable, let's age as well as possible so we can enjoy a long, healthy life."

"This kind of test," I press on, maybe to delay the moment when they'll take my blood, as I'm not sure I actually want that to happen, "it was invented here, and this is the only place you do it?"

"Yes, it's the result of many years' research."

"And it hasn't been patented?"

"That's not possible, it's not patentable. It can be registered, at most. What we do is publish our method and our results. It's only done here in the Complutense. Not outside Spain, either, not especially because of the cost but because it's very painstaking work — you need specific expertise."

And then, I don't know whether voluntarily or out of social pressure, I submit to the extraction of blood, and I think they take out more than usual, just as they do to the palaeontologist.

When we leave the faculty, it's still early, and the cool morning air helps restore me to myself.

"Are you sure about what we've done?" I ask Arsuaga.

"Totally!" he exclaims with the kind of optimism he has on his optimistic days.

It's 29 June. The following day, the palaeontologist will go off to his excavations at Atapuerca, and I will retreat to my house in Asturias till September. Mónica de la Fuente has said she'll email us our results in a couple of weeks.

I spent those two weeks rather on edge. I was unsettled by the fact that there might be a discrepancy between my biological age and my chronological age, as that was one more dichotomy to add to a whole big heap of divisions that entailed a view of the world that was troublingly dual. In this heap there writhed around, like vipers in a nest: essence and existence, life and death, waking and sleep, reality and fiction, madness and sense, body and mind, youth and age — not to mention up and down, inside and out, left and right, mere adverbs of place that are nonetheless endowed with an intolerable moral overtone.

Chronological age / biological age.

Oh God.

I remember that on one of those days, having come back from buying the paper at the square in Muros de Nalón, the municipality in Asturias to which I'd withdrawn, I walked past a house in the doorway of which two women were talking. One of them was saying at just that moment, "Paco died — Fina's Paco."

"Rosario told me," said the other.

A death was reduced to this, a banal conversation on the street.

Paco died — Fina's Paco.

I also recalled the exclamation of the protagonist of *The Death of Ivan Ilyich* at the very moment of his passing: "So that's what it was!"

I always interpreted this as Tolstoy's character wanting to say that it was all just a trifle. That death was some silly thing. That maybe dying wasn't even dying, but changing state, just as water doesn't die when it evaporates or is transformed into ice. A shift within life.

So that's what it was!

And so what had life been?

I recited those terrible lines of Idea Vilariño's to myself:

What was life
What
What mouldering apple
What dregs
What waste ...

Finally, one morning, when I was checking my emails, a message popped up from Mónica de la Fuente with a document attached. I hurried to open it. I saw some graphs, some colours, some arrows, some percentages, but my eyes ran down the whole report in search of the final conclusion, if it had one.

And it did. It said: "BIOLOGICAL AGE: 50."

To this bare piece of information, she had added the following NOTE: "Congratulations, you're ageing at the speed one would expect to find in a 50-year-old."

I couldn't believe it. The discrepancy between this and my chronological age was no less than twenty-five years, a whole quarter-century in my favour.

Ecstatic, I sent the report to Arsuaga, who in turn let me have his. I scanned it, anxiously, till I found his BIOLOGICAL AGE: 65.

In his case, then, there was no discrepancy at all between chronological appearance and biological reality. The palaeontologist remarked in his message:

Millás, great news! I'm much older than you. A
most succulent literary subject for you, I think.
Abrazos

Far from being a "succulent literary subject", it made me feel sad. Was something going to happen to the palaeontologist, I wondered? That physical optimism that was so characteristic of him, those bursts of militant Epicureanism, might they not be there to help hide, or deny, some kind of psychic despondency that, either out of shame or panic, he could never let out? I remembered those moments (fleeting ones, in truth) when Arsuaga would disappear into himself as if in a kind of Zen exercise, and I subjected them, in the light of Mónica de la Fuente's report, to a totally different reading to the one I'd given them previously. Maybe he was at his very lowest, maybe it was only a momentary depression, from which he would propel himself back to the surface again, feigning that spirited, tireless guy that perhaps wasn't him at all.

I felt disconcerted in the extreme. I couldn't answer his message.

A few days later, he wrote to me again. He said:

> Millás, there's an important nuance in all of this that you ought to raise with Mónica de la Fuente — it'll add to the discussion. What the tests show is not that you have the body of a fifty-year-old, but rather that you are ageing at the same rate as the average fifty-year-old. I'm sixty-six, but I'm ageing at a faster rate than you, which means, in short, that I might die after you and come to your funeral, and yet I would still die, though at a later date, at a younger age. Do you see what I mean? This idea of biological age is poorly understood. What's being measured is the pace at which one ages. I ought to slow down the pace of my ageing, and you, now that we're about it, could be doing the same. In any case, congratulations! I don't mind us dying together.
>
> Abrazos,
>
> Juan Luis

Reading that was a liberation. Arsuaga was rationalising the subject and returning to his usual mood. I'd got used to a palaeontologist who was lively and a Millás who was listless, and the role reversal had hit me hard.

I wrote to the doctor asking the suggested question. Her reply was as follows:

> That's a very good question ☺
>
> I'll answer you with a fact that we have discovered (and published on). Those centenarians we've tested had a biological age much lower than their chrono-logical age (around sixty, but there were even some at forty). Do you think that despite this biological age they could go off dancing all evening, the way an actual forty-year-old might? No. However, they were well, in general, and they had reached a hundred because their ageing was happening slowly and they'd had, and still did have, good mechanisms for adapting.
>
> Let's keep the conversation going.
>
> love
>
> Mónica de la Fuente

The palaeontologist, to whom I forwarded it, agreed with this nuance. Then he added, in another message:

> We all have our challenges, Millás, and mine is touching my toes without bending my knees, and attempting this both when sitting and when standing. They say elasticity is health and long life, which makes me laugh because it reminds

me of that Buffon theory I told you about, the
one according to which, as with plants, getting
old means growing tough and woody, while
youthfulness is about being flexible as a reed. I'm
still a long way from touching my toes because I
spent so many years lignifying, but I hope to get
there one day. Of course, women can do it with
no problem, but they don't lose elasticity the way
we do, what with our muscles getting shorter. I'm
thinking of practising on the beach. As I say, we all
have our challenges.

An abrazo,

Juan Luis

To me, this elastic Arsuaga was one hundred per cent pure
Arsuaga, and that calmed me down for good.

In September, if we were still alive, we'd see each other
again.

ELEVEN

The tree man

In early August, I received the following message from the palaeontologist:

Dear Millás,
Touching my toes is impossible. Nonetheless, I'm almost there with the sit-and-rise thing, because I'm already used to sitting cross-legged on the ground for my excavations, and having to get up from this, to me, familiar position is something I've been doing for many years — including this July, in fact: I can still do it. But I am lignifying and I haven't managed to get even one millimetre closer to touching my toes, they might as well be at infinity — and beyond. I'm a tree man, an old trunk, dry wood, sure to be felled by the next storm to come along. Remember this, because we're going to talk about tree trunks on our next outings. I hope you're doing well, relaxing, getting lots of work done. You came to mind today because I went to the local market (El Puerto de Santa María) to buy some sardines for a barbecue,

and ended up gorging myself on churros — they
make really excellent ones at the La Charo stall.

    A big abrazo,
    Juan Luis

Such an explicit recognition of his physical limitations did
worry me a little, but I like the image of lignification, which
inevitably reminded me of the lines from Antonio Machado:

> From the old elm, cleft by the lightning-ray
> right down through its rotten core,
> with the rains of April and the sun of May,
> green leaves have emerged once more.

I sent them to the palaeontologist. Perhaps we, similarly
cleft by the lightning bolt of the years (each decade a
thunderclap), might yet have a few green leaves left to emerge.
This is how Machado's poem ends:

> My heart waits too,
> towards light and life,
> another miracle of spring.

Incidentally, by "sit-and-rise thing", Arsuaga meant an
exercise that consisted of sitting cross-legged on the floor and
getting to your feet without supporting yourself on anything,
which we'd done in the Pilates class. To expand my knowledge
of it, I had resorted to the internet, where I'd found a sitting-
rising test in which they measured your likelihood of dying in
the next six years according to how well you were able to carry
out this practice.

I didn't try it myself, but I remembered that Corbalán had recommended a strange physical exercise for our age that involved throwing yourself to the floor and getting back up in any way that occurred to you. I reminded the palaeontologist of this, and I tried it myself regularly over the summer: I would lie down and get back up resting my hands on the floor or on a chair, though not without difficulty. I liked doing it for the exercise's metaphorical dimension, which seemed somehow moral in nature.

Being prepared to fall.

Being able to get up.

In a way, isn't that what life is?

Arsuaga replied to my email unusually quickly:

> Dear Millás,
> We can use the end of the Machado poem for
> what we're moving onto next in the book, so hold
> onto it for that. And indeed, if we did sit on the
> ground more often, to play with our grandchildren
> (I don't have any), for example, or to dig, we'd be far
> more flexible and far less lignified than spending
> all day sitting writing books on the computer,
> barely even crouching down once in a whole day
> to pick something up. Which brings us back to the
> Palaeolithic niche again: sitting on the ground with
> the children, being down on our haunches when
> we talk to our friends, digging for geophytes, et
> cetera. I don't know if we'd have longer lives in that
> case, but flexibility is a luxury enjoyed by almost all
> women and, sad to say, very few men. Might this
> have something to do with the fact men don't live

as long? So much testosterone, so many muscles that are bulky but short, so much exhibitionism — we were always going to have to pay some price when we got older, that's what old Williams would have had to say about it.

By the way, the latest religion in gyms is the plank, which is an exercise you or anyone could do. You can set yourself little challenges: holding it for fifteen seconds to start with, moving on to thirty seconds, and finally a whole minute. They're called "isometric exercises" because they don't shorten the muscles, they only maintain their tension and length.

I'll also confess that I was very surprised you hadn't put anything about basketball in the text you sent me, because that was my definitive weapon against Lamarckism, plus it was no easy thing getting us into that training session — given Covid, and given how late into the season it was, not to mention a number of other factors. I'm wondering if you understood why we went. The basketball players weren't tall because they played basketball (which would be Lamarck's take), but rather, they played basketball because they were tall, which was down to heredity (Darwinism). Which was the exact same reason I took you to see the giraffe in the Museum of Natural Sciences beforehand, that animal being the example Lamarck gave of acquired characteristics. I'll run you through it again: according to Lamarck, giraffes have long necks because their predecessors put so much effort into reaching up for the leaves in the treetops, exactly like me trying to touch my toes, millimetre by millimetre, making tiny gains but also losses every day, painstakingly — like a giraffe with a short neck that can't get to the leaves on the trees and ends up dying of starvation. But the information goes from the genes to the phenotype, not from the phenotype (the body) to the genes. Nothing you do during your life will modify your genes. Besides, I've passed mine on to my three children already.

If you remember, there was that one girl we were talking to during the training session, who told us she had very tall parents, and that was why she was tall (and not because she'd been playing basketball since she was little — she laughed when I put the idea to her, the suggestion she'd grown so tall through playing basketball all her life — to her, the question seemed ridiculous), and I later got her to tell us that she'd never marry a man who was shorter than her, not a chance! In other words, she'd only get married to a man who was taller than her, or the same height, which meant her children would be very tall as well, and they'd get them playing basketball from a young age and turn them into great champions, who in due course would have their own children, who would be just as tall as their parents, if not taller …

This is the best example I know of the way in which evolution functions, and of why Lamarck was wrong and Darwin was right. It took a hell of a lot of work to get us into that training session. I

planned the visit to the museum to see the giraffe
as a way in to Lamarckism, which I was going to
sum up by talking about basketball — and then you
didn't even mention basketball …

Is it because you're allergic to sport in general,
in spite of all the lessons it has to offer? That's
two stories from sport that haven't made it in: in
the first book, the bit about football and symbolic
exclusive identities, and in the second, basketball
and Lamarkism.

An abrazo,
Arsuaga

The palaeontologist was referring, here, to a football
match I had refused to watch because crowds scare me, and
to a subsequent meeting, which I have not recounted, with a
women's basketball team who were training at one of the courts
at the Complutense. Apparently, he'd had to ask several favours
to get us in to talk to one of the players, Paula Real. We asked
her why she played basketball, and she said it was a sport she'd
played her whole life.

"I've given others a try," she added, "but none of them gave
me as much fulfilment. I'm happy when I'm playing it."

"Are your parents as tall as you are?" asked Arsuaga.

"Yes, my dad's 1.95 metres, my mum's 1.77."

"And are you this tall because you've been playing basketball
since you were little?" Arsuaga pressed her.

"No!" replied the young woman, smiling.

"Oh, Lamarck!" proffered the palaeontologist. "So you're
saying you'd have grown the same amount even if you hadn't
played basketball?"

"Of course!"

We visited the court at the Complutense on the same day we went to the Museum of Natural Sciences to see, among other animals, the giraffe. Arsuaga does nothing without a good reason: the giraffe didn't have a long neck from stretching it up to the treetops, and Paula Real wasn't tall just because she played basketball.

"Another blow to Lamarck, and a big one," the palaeontologist said to me, smiling, when we had left the court.

And back in the car, just in case, he rounded off the idea: "Write this down, Millás, it's very important: it's the sport that selects the biotypes, the morphologies, not the other way around. Tall people are also really good at long jump, because of their long legs, and because they run a lot. When you arrive at the take-off board, you've got to be running fast. In the gym, on the other hand, everyone's short. In football, you get all kinds of biotypes, except for the goalkeepers. Got it?"

"Got it," I said.

So now I know.

TWELVE

To hell with them

Once we had rounded the Cape Horn of the month of August, I replied to a brief email from the palaeontologist in which he was asking me to perform a task that was impossible to carry out in my situation. I told him as much:

> Dear Arsuaga,
>
> I'm very sorry I couldn't watch the gulls and make a note of their habits as you asked me in your last mail. I've had relatives and friends staying and they've taken up all my time these late August days. To make up for it, I have noticed a domestic incident that has rather disturbed me. You see, here at the summer house we use butane gas both for cooking and for the water. Normally, up till now, whenever they bring a replacement cylinder, I've always asked the delivery guy to leave it near the house in an outbuilding that was once a stable. This year, though, I asked that he bring it right to the kitchen, for fear I might not be able to lug it over from the stable myself when the current one

runs out. I think this suggests a somewhat greater acceptance of my limitations. I am indeed old, and everything around me reminds me of that fact.

I wonder whether the people who decide on the weight and size of these cylinders think about the typology of man or woman capable of transporting these big orange things from one part of the home to the other. If they do, am I now excluded from the group? Should I give up using that sort of energy to meet my domestic needs?

Last year, I renewed my ID card, and they gave me one that expires on 1 January 9999. Yes, you read that right: the year 9999. I will expire before then myself, of course. The Ministry of the Interior issues this crazy document after the taxpayer hits seventy (the longevity of our species, incidentally). Meaning that they consider him written-off. Another exclusion, then. The tribe, one way or other, is pointing me the way to the graveyard. But despite my seventy-five years, I do still need gas and an identity. I'm not ready to give up either one.

But they could produce gas cylinders that are lighter, and they could renew our ID for a given time so as to give us a bit of life expectancy. The exclusion from the tribe (to which, honestly, I've never felt all that united, and I'm starting to see why) is cruel and is intended ultimately to make us voluntarily exclude ourselves. The implicit order is as follows: off you go to live on the sofa, tune in to the tackiest program on TV, surrender yourself for a time to mental and physical decay, and then die.

Well, to hell with them.
I have no intention of doing that.
Abrazos,
Juanjo Millás

The palaeontologist answered right away, but his reply was disappointing:

Dear Millás,
Don't worry about the gulls, but when you're on the
beach, do check them out. You can even give them
a bit of food if there's nobody around: watch and
see if they compete or cooperate.
Over the years, the exact same thing happens
with architectural barriers as happens with mobility.
It isn't until you break a leg and have to go around
in a wheelchair that you realise everything's geared
towards the young. You find there's absolutely
nothing you can do, not even the simplest things,
because everything's pitted against you.
Abrazos from Pinilla,
Arsuaga

On the one hand, he was telling me not to worry about the gulls, and on the other, that I still shouldn't not watch them. That's what obsessive temperaments are like. He despatched the rest of my message in a few swift sentences. I forgave him because at the time he was already working at the Pinilla del Valle site, where the Neanderthals of our previous book were, but his offhandedness did wound me, though I preferred not to tell him. Did he actually have any idea what it was like to carry a butane cylinder or to have an ID document in your wallet that no border guard would ever believe was genuine?

THIRTEEN

The secret life

On 31 August, first thing in the morning, I'm packing my suitcase to return to Madrid after my Asturian holiday, when my phone rings.

"Done your densitometry yet?" asks Arsuaga point-blank.

"No," I say, "and I don't intend to. I'm not going to any more doctors and I'm not doing any more tests. You seem determined for them to find something wrong with me."

"You'll see it for yourself," he says, "but densitometry is important in finding out the state the skeleton's in. In old age, the bones become softer, they decalcify, and osteoporosis starts to set in, while at the same time the arteries calcify, giving rise to coronary disease — arteriosclerosis. It's like the calcium's being transferred from the bones, which it's good for, to the arteries, where it does all kinds of mischief. Remember the theory of pleiotropic antagonism? We talked about it the day we went to the gym."

"Of course: according to that theory, the same gene that helps keep our skeleton in shape when we're young is responsible for our arteries hardening when we're older."

"There you have it. Will you or won't you do a densitometry test, then?"

"I won't. Anyway, I've got my bags packed. I go back to Madrid this afternoon."

"I'm at the dig in Pinilla, we're on a break. I sat down on a rock and you came to mind."

"Right," I say, hoping to curtail the conversation.

"Are you standing up?" he asks.

"I am."

"Well, sit down for a second. I've got something I want to tell you."

I sit down, resigned, on the edge of the bed, next to the suitcase, which is spread-eagled, with three or four disordered shirts, that look like guts, inside it.

"I'm all ears," I surrender. "But just don't keep insisting about the densitometry thing."

"No problem, but tell me this: what is it exactly that we're looking for?"

"I've honestly never known," I reply.

"The reduction in infant mortality," he goes on, quite oblivious to my aggressiveness, "has brought about an astonishing increase in life expectancy. More people live to a post-reproductive age than ever before. The idea of meeting your great-grandchildren, of getting to play with them, is no longer utopian. Transmissible diseases are combated with vaccines, with antibiotics and antivirals …"

While he talks, I imagine him on the rock he mentioned, dressed like Indiana Jones, watching the river flowing by at the bottom of the valley.

"And if you're involved in a car crash," I hear him say now, "the traumatologists can get hold of you, and you end up good as new. We've made great strides when it comes to biomechanics. And cataracts don't mean going blind any longer. Can you imagine what it would been like to break your femur in the Palaeolithic?"

Arsuaga is talking slowly, as if he's taken a sedative or as if he's suffering an attack of nostalgia.

"Are you feeling nostalgic?" I say.

"Nostalgic for what?"

"I don't know, you sound funny."

"I'm tired, it's been a lot of work here. But I wanted you to make a note of the fact that, as well as prostheses and other solutions that lead to longer lives, another avenue is that of cellular research."

"The secret life of cells," I say, repeating a syntagm that he says often.

"There, in that secret life, we might be able to uncover the mystery of ageing, whatever that actually is."

"'Whatever it is'?" I say. "You mean we're still trying to figure out what it is?"

"Oh," he replies, unperturbed, "there's no doubt how helpful it would be in tackling this subject to have a scientific definition of this biological process which, don't forget, is something that only humans suffer from, along with our pets and those animals we keep in zoos. The ones in zoos experience a kind of decline at the end of their lives unknown to any of their predecessors over the hundreds of thousands of years of their species' existence. But it isn't enough to see ageing as purely a case of the loss of faculties we associate with old age. The most precise definition of ageing we currently have is that of someone's — anyone's — probability of dying before the year is out."

"Well," I proffer, "eight months of this year have already gone by and we've survived. Besides, we've got to deliver the book in December, and I've never defaulted on a contract. So you can relax."

The palaeontologist gives a little cough. Now I imagine him gazing off at a point somewhere high in the mountains. It's a shame he's not a smoker, I think: a smouldering cigarette between his fingers would suit him right now.

"Mortality," he continues, "is very high in all animal species early on in life. After that, it decreases to its lowest level. Then, from the moment they hit the age of being able to reproduce, the probability of death doubles every certain number of years."

"And in the human species?" I ask.

"In the human species," he says, "it doubles every eight and a half years." *Eight and a half*, I repeat to myself, making an automatic association with the Fellini masterpiece. "The pace is almost the same in African elephants. With mice, it's every four months. With dogs, every three years."

"They've tested that?"

"Yes. It's known as Gompertz's law, after Benjamin Gompertz, who came up with it in 1825."

"Who was he?"

"An actuary."

"So the insurance companies know when we're going to die."

"Of course, which is why they always win, just like the banks. So, in species like rabbits that have short lives, the probability of death is already very high when they reach adulthood."

"And on top of that," I emphasise, "it keeps doubling every so often."

At that moment, my wife comes into the room and gesticulates to ask how I'm doing with the suitcase. I signal back to demonstrate my powerlessness, and tell her, covering the microphone, that I'm talking to Arsuaga.

"Well, the cat's got away!" she exclaims.

I tell the palaeontologist to excuse me for a moment, and I turn back to my wife.

"What?" I ask, alarmed.

"I said the cat's escaped."

"But he was already in his cage."

"But he opened it somehow and he's gone."

At our house in Asturias, the cat spends his whole life outdoors, but on the day we're leaving we put him into his cage first thing in the morning so he won't go out, as you can never tell when he'll come back. We can't leave without him, but we also can't delay our trip, because my wife and I both have work tomorrow in Madrid.

"I'm going to call him, see if he shows up," she says, leaving the bedroom.

"Is something wrong?" asks Arsuaga.

"Our cat's escaped and we've got to get back to Madrid today."

"I already told you cats were domestic, but also not really."

"It's going to mess everything up for us," I grumble.

"With lynxes, on the other hand," he continues, oblivious to my domestic woes, "which, by the way, are one of your pet's predecessors, the probability of them dying when they reach adulthood is already lower than that of rabbits. It's not for nothing that they're the kings of our Mediterranean ecosystems."

"I don't know where you're headed with all this," I say, praying to every power that the cat reappears and that Arsuaga has the tact to say goodbye.

"In that case, I'll go back to bodily cells," he says decisively. "The free radicals in oxygen are very dangerous because, since they contain one unpaired electron (it's all alone, poor thing), they're able to react with other molecules and cause them damage, especially in the mitochondria, which are the organelles where the cell's energy is produced."

"In other words," I say, "we're talking about what in other conversations we've called oxidation."

"Exactly," says the palaeontologist. "As a way of reducing the oxidative stress caused by the hateful free radicals, the cells produce antioxidants."

"Which is why," I interrupt him, "I take melatonin."

"Okay. So, you'll remember that oxidation can mean species with high metabolisms not living very long, because they burn out. That's the rock-star theory again: 'live fast, die young, leave a good-looking corpse'. These species would accumulate, on account of their frenetic lives, lots of free radicals in a short space of time, lots of oxidation, and they'd pay a heavy price for it. Whereas animals with low metabolisms, with fewer heartbeats per minute — like elephants — would accumulate fewer free radicals on a yearly basis and, as a consequence, live longer."

"Size equals duration," I summarise.

"The larger the animal," Arsuaga continues, "the more cells in its body, and so the more cellular division there is taking place on a daily basis, which necessarily means more random mutations and therefore what should be a higher probability of cells mutating, becoming cancerous. And yet that isn't the case: generally, larger animals live longer than small ones. They somehow manage to delay the appearance of their tumours. There's still a great deal we don't know."

"And cells?" I ask, while simultaneously listening to my

wife's voice as she calls the cat from the garden. "Can they keep dividing indefinitely?"

"No. There's something we haven't got onto yet: namely telomeres, which elongate the part of the DNA that's at the end of the chromosomes."

"Kind of like those things on the end of shoelaces?" I ask, always on the lookout for a clarifying image.

"Kind of like that," says the palaeontologist. "The thing is, with every cellular division, the telomeres are cut shorter, like they've been snipped with scissors. After lots of divisions, they've been cut so much shorter that the cell can't divide any further."

"So is longevity related to the length of the telomeres?"

"Exactly. If we found a way to repair these telomeres, the cells would go on dividing forever. There's an enzyme, telomerase, that repairs them, but the cells in human bodies, unlike those in mice, don't produce it, meaning that whatever happens, the number of divisions a cell can undergo is finite."

"Why is it that the cells in human bodies don't produce that enzyme, this telomerase, and the ones in mice do?"

"Because telomerase is a double-edged sword. Just so we're clear, the cells in tumours *do* produce telomerase. And because of that, they're immortal. You could cultivate a tumour in a lab indefinitely. A cell from human tissue can divide fifty or so times over the course of an adult life. After that, you don't get any more renewal of the cells."

"You get old age," I conclude.

"Indeed."

"But if science could repair the telomeres …"

"In that case, sooner or later, the mutations would appear — that's to say, the cancers."

"So we're trapped: if the lack of telomeres doesn't kill us, cancer will."

"Well, the theory still needs refining because not all the facts we have actually fit. For example, mice have far longer telomeres than we do, and yet they have these short lives. In any case, the study of telomeres is a hugely important research field in cellular biology, and therefore in everything to do with ageing and cancer. A better understanding of the immune system is also very important: that's the only way we'll detect and eliminate the mutated cells before they proliferate and spread throughout the entire body."

My wife reappears and makes a helpless gesture. No sign of the cat. Then she uses her index and middle fingers to make the gesture imitating the movement of scissors — cut off the conversation, because this is neither the time nor the place. I nod a yes to her, but Arsuaga is really on a roll: it isn't easy finding the moment to hang up. And in fact, when my wife leaves the bedroom, instead of hastening the conversation along, like an idiot I ask him, "And do all the cells in the body divide?"

"Not all of them. The neurons, for example, barely divide, and the same goes for cells in the heart. Everything that happens at a cellular level is very interesting, but very complex. For now, it's enough that we've broached the subject, but in general we've been looking at the question of ageing with a wider focus than the molecules and cells. That is, we've been looking more at the level of the individual and the species. That's why we've gone so deep into the subject of the Palaeolithic diet and niche, which, indirectly, lead us to autophagy and cellular cleanliness, remember?"

I get the impression the palaeontologist is pausing on our

path to cast a glance backwards, to what we've already studied, sort of by way of summary, while also informing me, without making it explicit, that we are coming towards the end of our work. The idea of finishing makes me feel nostalgic all of a sudden, maybe the same nostalgia that is afflicting him, sitting there on his rock, in the middle of the Lozoya river valley, the Valley of Silence. It seems to me, also, that he is trying to tell me, subliminally, that if we don't go deeper into the cell question, it is only because of my cultural inadequacies. And he is right: on the free radicals, and on the telomeres, and on the telomerase, I've had to make a superhuman effort to understand what he is trying to explain. I thank him inwardly for presenting the question so tactfully. The palaeontologist is usually very tough on other people's ignorance, but he does occasionally take pity on the specific ignorance of those who, like me, show some interest in learning.

"So?" I say, to encourage him to go on.

"So," he concludes, "healthy living, daily exercise, eliminating stress from our lives, a decent night's sleep, good diet, no smoking, no drugs, and don't stuff our faces constantly throughout the day, rather only when we deserve it."

"In short," I say, "Epicureanism."

"At some point," he adds, "we ought to talk about the difference between Epicureanism and hedonism, because people get them mixed up when in fact they're completely different. We're currently witnessing a pandemic of people with weight problems and pathological obesity, which has more to do with hedonism than Epicureanism."

"Right."

"In terms of what you and I are concerned with, the question we're asking is why there are some species that live very short lives and others that live very long lives. We've tended to use the binoculars rather than the microscope until now, and that's actually what we're going to carry on doing the next few times we meet."

At that moment, I hear a noise at the window and see the cat, who is standing there looking at me suspiciously. The open suitcase signifies travelling to him, and travelling means the cage. I pretend not to see him, so that he feels more trusting, in the hope that he'll come closer and I can give chase.

"Keep talking," I tell Arsuaga.

"What?"

"I said keep talking. My cat's just appeared and I need to act natural so that he'll come closer."

"Okay … I don't know what to say."

"Which is pretty weird," I complain. "On the whole, you don't even shut up underwater."

"That is a problem I have," he admits. "I go quiet when people need me most."

"Well, do me a favour, make an effort, and just keep talking. If you can't think of anything, recite the periodic table."

"Okay, Millás," he continues, clearing his throat a little, "so, we'd all be delighted if we found out that every species has a biological clock inside its cells, a clock that determines the pace of the different ages of life: being embryos, fetuses, babies, children, preadolescents, adolescents, parents, being in the menopause, being grandparents, great-grandparents — and finally dying. And we like the idea because, if such a clock existed and we got our hands on it, possibly we'd be able to stop it, and that way become eternal. If death were pre-programmed, all we'd have to do would be to de-program it."

"We'd have a cure," I say, surreptitiously watching the

movements of the cat, who has come into the bedroom and is advancing with the greatest caution towards the bed on which I am still sitting, beside the suitcase.

"Shall I go on?" asks Arsuaga.

"Of course. I think it's dropping its guard."

"The alternative to the theory of pre-programmed death," he continues, "is the Medawar/Williams theory, which we've covered. Neither death nor old age are pre-programmed, according to them. Each is simply the result of the accumulation of lots of different mutations that natural selection hasn't managed to eliminate over the course of our evolutionary history because they're expressed so late, at a point when almost nobody continues living, meaning they aren't on its radar. As well as being the price one has to pay for being so godly in one's youth."

"There's no hope for immortality in that theory," I point out.

"None at all," he says. "Going okay with the cat?"

"You're not going to believe it!" I exclaim. "He just got onto the bed and climbed right into the suitcase, without any fuss. It's as though he's surrendered. But also like he's adamant that he will *not* travel in his cage, and we'll have to take him in the suitcase instead."

FOURTEEN

Shangri-la

The month of September awoke all of a sudden, without bothering to stretch, as if there were no time to lose. On Friday the 10th, Arsuaga asked me to meet him in Rascafría, a small town in the Madrid mountains close to the Pinilla del Valle Neanderthal site. I let myself be guided by my car's sat nav, which chose the longest and surely the most tortuous route, as it took me via Soto del Real and Miraflores instead of along the Burgos highway, which meant I had to go over the Puerto de la Morcuera, winding this way and that, and as you came around the bends it wasn't unusual to run into cows placidly crossing the narrow road in search of the grass on the other side. The grass on the other side is always better.

Over the summer, the palaeontologist had been investigating whether one of those mythical valleys you get in TV news bulletins actually existed, those ones whose inhabitants live longer than usual. But he didn't find it and so we couldn't visit it. There were just groups of centenarians in old people's homes, facilities that didn't correspond with the idea we had of outdoor spaces. However, he told me that in the Rascafría forests he had met an extremely interesting lady who was over a hundred ("quite a lot over," he added), lived alone, and would be delighted to receive us and perhaps reveal to us the secrets of her longevity.

I arrived at the place we were to meet at 9.00 a.m. There was that coolness with which September warns of the impending October, but I was reasonably well bundled up — though irrationally clothed, to judge by the palaeontologist's expression.

"I warned you we'd be heading into the forest," he said reproachfully, proffering his hand.

"I'm still waiting for you to take me to Decathlon," was all I said.

The palaeontologist had lost weight in the two months since we'd seen each other last, his colouring was good, and he'd just had his hair cut. All this gave him that aged adolescent look that works so well for him and provokes such envy in me.

He was waiting for me with a man to whom he introduced me straightaway: José Antonio Vallejo, the local "forest-keeper", according to the old terminology. Vallejo, who knew the centenarian old lady, would guide us to her, since her house, from what I was able to deduce, was not very easily accessible.

"She lives on her own in the middle of the forest?" I asked.

"Yes, in a kind of hut. We forest-keepers visit once or twice a week in case she needs anything," he replied.

We got into the car that the palaeontologist used on his excavations and which wasn't a car as such, but an off-road Santana Aníbal — or Hannibal. It was like some kind of mythical Land Rover I'd never heard of, sort of a cross between a rhino and a tank — pure biotechnology, in other words — but I didn't dare say so for fear of putting my foot in it, as Arsuaga seemed quite in love with the thing. He even made me

crouch down to see the underside and check with my own eyes
that, in addition to shock absorbers, it had coil springs, devices
more suited, I thought, to ancient stagecoaches than modern
vehicles.

As if he'd read my mind, the palaeontologist said, "Springs
revolutionised carriages. They must have been invented in the
eighteenth century or thereabouts. This car is the latest thing in
the history of springs — you could scale walls with it."

What was certainly true was that that hulking thing,
which was maroon-coloured, had a kind of gross beauty — a
pre-tax beauty, you might say. Arsuaga got behind the wheel
and Vallejo into the passenger seat. I occupied the seat behind
them, which was hard and uncomfortable and rather like being
in barracks: in short, hostile. The interior had dry mud stains
here and there.

"We're going to the Barondillo hill," said the forest-keeper,
who knew the place like the back of his hand, as he'd been
looking after it for twenty years. "We'll get as far as we can by
car and then we'll walk, though you're not exactly wearing the
most suitable shoes."

"Right," I admitted, hanging my head.

But then I quickly recovered and asked, "How's it possible
that a woman aged more than a hundred lives someplace so
isolated?"

"That's precisely why we've come," said Arsuaga, "to see this
unusual thing that'll be helpful for our research."

The first stretch of our trip, in spite of the discomforts of
the vehicle, turned out to be bearable, but then we entered a
dense forest whose floor was irregular in the extreme, with
potholes, with bumps and recesses and rocks like watermelons,
so that the vehicle jumped about like a little canoe in the

middle of a huge swell. There weren't enough handholds inside the car to maintain one's posture or poise. Arsuaga, gripping the steering wheel as if he was helming a boat in the middle of a storm, laughed each time there was one of these jumps, proud of the excavation site's Aníbal. Even now, as I write these lines, I can feel that upset stomach I experienced then and which I didn't dare to reveal, though it must have been visible in my expression, which the palaeontologist was watching in the rear-view mirror, fearful, I thought, that I might throw up inside that sacred space.

Soon we reached an earth track that was barely more agreeable than the previous stretch; and at times, much less so. On both sides of that kind of firewall, there rose up to the sky, spear-like, aged pines, some of them with their branches truncated, I imagined, by the inclement weather and the weight of the winter snows. Then, to our right, a river appeared — the Lozoya, I was told — and we were seeing its upper reaches, known as the Narrows. I asked, so as to appear interested in the things of this world, what altitude we were at, and I was told, "around sixteen hundred metres".

"This is where the springs really show their mettle," said Arsuaga. "I don't know if there's a modern shock absorber that could deal with this."

"There must be," I said. "Seeing as they do make them, and they work."

"We'll have a brief stop here," was all he said, which I thought was to allow me to recover from my travel sickness.

"We're going to see the Juanón pool," added the forest-keeper.

Since my only experience of forests as thick, dense, variegated, shadowy, and mysterious as this one were from the illustrations in the tales that were read to me as a child and that I read in turn to my children, I had the impression of being inside one of those children's stories, more than an actual forest.

"Look at the blackberries," said Arsuaga, pointing into the bushes. "Most of them are still green, because everything here's a month behind. The forest fruits mean the animals can survive through to spring."

The ground looked like it was also carpeted with somewhat disturbing purple flowers, with barely any stems, whose petals opened into the air like the fingers of a hand to show their complex venereal nether regions to all and sundry. I asked what they were called.

"It's an autumn crocus," said Arsuaga, "or 'picnic's end'. It's called that because it comes out at this time of year, when it's getting colder and people stop coming for picnics in the forest."

"In other places," added Vallejo, "they call it 'scare shepherd', because this is also the time when shepherds look for other places to graze their sheep. It's a type of saffron."

The Juanón pool was a hollow in which the water from the river collected, creating an emerald-green stain that looked like it had been put there not so much by nature as by an interior decorator. That was the problem, I thought, with knowing the representation of things, as I had done, before encountering the things themselves. I took the landscape to be a copy of the illustrations from fairytales rather than taking the illustrations from fairytales to be copies of the landscape.

Contributing to the feeling of unreality to which I'd fallen victim, I discovered that, just before the pool came into view, a lovely medieval bridge rose over the riverbed, its stones worn by the passing of time and exposure to the elements, and covered in moss; a bridge, in short, that though it did enjoy the three dimensions that are the property of real objects, rather evoked the work of a romantic painter than of a practical architect.

Making the excuse that I needed to urinate, I withdrew behind some bushes and washed my face in the river in order to recover from the car sickness. The water was cold and as transparent as the air we were breathing. Once recovered, I looked at the forest in the hope that my previous feeling of unreality might have left me. Far from it; it was accentuated now, as I sensed that everything I could see was still a copy, a reproduction, an imitation. Then I recalled that idea of Oscar Wilde's: about how nature imitated art, which meant that I, in the middle of that facsimile of art, was surely a fake, too, a copy — but a fake or copy of what or whom? Where, then, was my true me? I thought about the story of Hansel and Gretel, which could easily have taken place in this strange spot, and wondered whether the old lady we were planning to visit also looked like the witch from the famous Grimm tale.

Something happened, in any case, that frustrated me: I was unable to communicate with nature like this, so fertile (and ferocious). I didn't manage to commune with the water or with the shrubs or with the stones of the medieval bridge or with the insects that buzzed around the vegetation or with the birds that hopped from branch to branch the way obsessions hop from head to head. I addressed myself telepathically to all these elements, which ignored me completely, as if I didn't exist. Between nature and me, there was a wall of insuperable incomprehension, yet neither the palaeontologist nor the forest-keeper seemed to be suffering from it. There was this perversity in me, one I didn't wish to show off: the fact I was able to enjoy the description or the reproduction of a forest, but not the forest itself, which felt unreal to me because I also saw myself, there in its guts, as completely unreal.

I went back over to my hosts, who enlightened me about the kind of rock that was prevalent in the area and that was not granite, as I'd believed, but a metamorphic rock, so called because it was the result of the transformation of another that, deep in the earth, had been exposed to pressure, heat, and a number of chemical agents.

"Look at the willows," I heard Arsuaga say.

And then: "See the heather."

Followed by: "Get a load of that bracken, it has such a prehistoric feel to it, and the birch, too, and the holly …"

The palaeontologist was naming the plants, and also the birds ("ah, look: a jay"), trying to convey a sense of communion with nature that I pretended to share.

Vallejo told me that the beautiful lichens hanging from some of the branches were called "druid's beards".

"Words just don't come near, right?" he said.

"Nowhere near," I agreed, once again remembering the illustrations from old stories featuring those prophets.

"Such a stunning day," said Arsuaga. "The air's so clean. Look at the leaves on that birch, the way they tremble in the breeze. This must have been what it was like in Paradise."

Back in the Aníbal, the powerful beast started to climb the hill, producing jolts and shudderings more suited to a fragile craft shaken by the waves than any land transport. The sickness came back, and I hid it as best I could.

We went up and up and up, as if climbing a wall of rough sea, while Arsuaga and Vallejo exchanged opinions about the wild fruit that appeared as we passed.

"Sloes," said Arsuaga, "don't turn sweet until after the first frost."

After some unspecified time, the vehicle stopped and we got out.

"From here," said the forest-keeper, "we go to the old lady's cottage on foot."

Arsuaga took my arm and invited me to distinguish individually between all the different things that appeared before our eyes.

"Don't see it as just a big, undifferentiated blotch," he said. "This, for example, is a wild rose, or briar, or indeed 'bottom cork', because it's an astringent, and animals use it to stop themselves dying of diarrhoea."

"How long to the old lady's cottage?" I asked.

"A hundred metres or so," replied Vallejo.

Very soon we stopped at a yew, of which this specimen, they told me, was the oldest in Spain.

"Between fifteen hundred and eighteen hundred years old," said the forest-keeper.

"She's a female," added Arsuaga. "She produces a red berry, as you can see, and the seeds are poisonous. The Cantabrians, Galicians, and Asturians who fought the Romans committed suicide with infusions of yew seeds, to avoid being taken prisoner. Strabo, the great geographer of antiquity, talks about it."

"The fruit can be eaten," said Vallejo, "but two seeds are all it takes to kill a person."

Incomprehensibly, the forest-keeper took a berry from the tree, put it into his mouth, and polished it off, skilfully spitting out its seeds.

"Want to try?" he offered me one. "They're really sweet."

"No!" I replied, appalled. "What if a seed slips through?"

I took a good look at the yew — its trunk (or, I should say, *her* trunk) was wide as a house, and very irregular, and hugely powerful thanks to the Herculean tendons running all the way up and down. The necrotic areas, far from conveying an impression of fragility, endowed it with a colossal vigour. From that trunk filled with organic shapes evoking forms of the aged human body, there emerged countless branches, some of them lignified, opening to a vast crown of leaves and fruit. The monster had a very human aspect, as it rested on the side of the mountain, in a place where the ground, which was very unstable, presented a stack of rocks to which the tree's roots clung like a giant's misshapen fingers in a horror story. At any moment, I thought, it might start to walk on those fingers, protected by a skin of moss, across the hillside, from rock to rock, in search of some corner where it could hide from our stares, in which I could detect just a touch of the obscene.

"The yew," said Arsuaga, "is the number-one sacred tree of the Celtic cultures. The wood makes excellent bows. By the way, it looks like the yew is clinging *to* the stones, but it's actually the yew that's holding them all together. If the tree dies, the substratum dies with it."

Worn out by these excesses of unreality, I turned around and asked the forest-keeper, "Well, so where's the centenarian old woman?"

Arsuaga and Vallejo laughed and replied in unison, "The yew's the old woman!"

I felt like a fool for not having guessed it earlier and for having believed naively that these sylvestral depths could be the solitary home to a woman of over a hundred.

"You said she lived in a kind of cottage," I said, in my defence.

Then they showed me, though I had seen it already, that the centenarian tree was protected by a fence, which, in a way, constituted her dwelling-space.

When we had arrived back, we ate (badly) in a local restaurant, where Arsuaga ordered snails to share, the portion of which turned out to be pretty stingy and whose sauce I found a little too spicy. Vallejo and I chose as our main a fish that they might have defrosted in a rush, because it was rather tough. The palaeontologist opted for some fried eggs with potatoes, which I watched enviously until they had disappeared from his plate.

"A corollary," he said, not long after we had sat down, as he used a toothpick to extract a snail from its shell. "Longevity, as much in the animal kingdom as in the world of plants, depends on the speed of maturation. Birch trees, willows, and poplars grow quickly, but they never live to over a hundred. Yew trees and sequoias, which are slow growers, can live to over a thousand. I was still a child pretty late into my life, so maybe I'm going to live a long time ... I'm joking, but write this down, Millás: those who grow quickly, die young."

"That we knew already," I said, bringing a piece of bread that seemed lately defrosted to my mouth. "The mouse and the elephant, the rock stars, and so on. What we're trying to do now is ascertain *why* there are species that live for centuries, what it is that causes that kind of longevity. And you haven't explained that to me yet."

"I have — when we ate at Zoko Retiro. We had caviar, remember?"

"I can still taste it."

"Well, even though it was only implicit, everything we're

going to go through now, I already explained when we were there. Some species are millenary, eternal, as much in the world of plants as in the animal kingdom, in the sense that they never stop growing, and the more they grow, the greater their reproductive capacity. And then there are species whose growth is strictly limited, like ours, in which, at a certain moment, growth stops, which coincides with the beginning of being able to reproduce; from then on, the reproductive capacity is maintained, but it doesn't increase. In reality, we see a decrease."

"That thing about there being species that never stop growing sounds like it's straight out of some fantasy story. Tell me about one of them."

"Well, the yew, for example, in the world of plants. And in the animal kingdom, the lobster, which we talked about at length when we were having the caviar."

"In that case," I replied, "there should be thousand-year-old lobsters out there the size of the Empire State Building."

"There would be, if it weren't for Medawar's test-tube theory. Don't you remember the waiter smashing the glass at Zoko Retiro?"

"Perfectly."

"He said that half the glasses would break within six months; after a year, there'd only be a quarter of them left; after a year and a half, only one-eighth, and so on successively. In the end, they'd all have disappeared, but due to external causes. I also talked to you at that same dinner about the fact that octopuses live to about two. It's curious, right? Then I asked for the caviar to be brought, so we could talk about sturgeon, which live to over a hundred and show no signs of ageing, though the unlucky ones do get caught in fishing nets. Pacific salmon, however, die the first time they spawn ... As you can see, there's a huge number of different longevities, and they all have their explanations."

"Are you coming full circle from that meal now?" I asked.

"What was it Chekhov said about a gun?" he asked in turn.

"He said that if a gun appears in the first act of a play, somebody must fire it in the second."

"Well," said the palaeontologist, "our gun was the lobster, which would live on and on, and always be young, if it weren't for a shark coming along and eating it. In other words: one can die from the inside out, like we do, or from the outside in, like lobsters or yew trees."

"I don't know …" I said doubtfully.

"That's only because you still haven't fully let go of common sense, which, when it comes to science, is the archenemy. I'll say it again: science is *counter*intuitive. And do me a favour, would you? When you're attempting to work something out, please try to suppress your 'what fors' — you resort to them a lot. There's no purpose to nature. Nothing happens for a reason. Everything's very subtle and very complex."

"And as for the lobster and the yew …" I said, to get the conversation back on track.

"They're both species that die because of external causes, the same way the glasses in restaurants break. And one other thing, in case I've failed to make it clear: why is it that in species whose growth is unlimited — which we're calling immortal — we don't get the expression of genes that produce deterioration, which we do in those whose growth is limited, like the species we belong to?"

"For that very reason, because they're immortal?" I ventured.

"Millás, that's a kind of circular thinking. Listen to me closely: they don't die because natural selection doesn't allow those genes to be expressed, and the genes of old age are never activated. Why is that?"

"You tell me."

"Okay: natural selection doesn't allow it because it sees that, although these individuals are ancient, they carry on reproducing in huge quantities, greater quantities, even, than when they were young. Understandably enough, among the thousands of lobsters that are born every year, fewer and fewer individuals remain over time, because they get eaten by a predator or because a wave comes along and smashes them against the rocks. For whatever reason. But the ones that *are* left make up for this loss with a greater reproductive capacity, so it all balances out in the end. Are you with me?"

"I think so. It's the same as what happens with the yew: you get progressively fewer generations over time, because a lightning bolt splits them in two or the earth opens up and swallows them, but the ones that are left make up for these losses by reproducing in greater numbers."

"It's complicated stuff, but you're getting it."

"So as long as you're reproducing healthily, natural selection doesn't bother you," I deduced.

"More or less. Now, this doesn't happen with species whose growth is limited, like ours. With every passing year, fewer individuals of your generation are left, but that isn't compensated by any increase in the offspring of those who survive. That's why you old folks become less and less important, as well as why you don't come up on natural selection's radar."

"In other words," I tried to sum up, "the greater the contribution of the individuals of a given age to the next generation, the more visible they are to natural selection, which will prevent them getting old. And so on, until there are very few individuals left of an advanced age who produce very few children in total."

"At which point," Arsuaga went on, "natural selection never gets sight of them. If we look after them so they don't die from the outside in, they grow old and die from the inside out, like domestic animals, animals in zoos, and us. And now, just to check you've really got this, let me ask you again: why are the genes responsible for strokes, diabetes, and Alzheimer's expressed in mortal species like the one we belong to?"

"This is becoming a familiar refrain: because at the age when they appear, we're supposed to be dead already, meaning we're not affected by natural selection, we don't come up on its radar."

"You've got it," concluded Arsuaga, dipping a piece of bread in the snail sauce.

I came back home via the Puerto de la Morcuera again, and I stopped at the highest point. It was starting to get dark in nature, but also in me, who belonged to a poor species of limited growth, the kind that dies from the inside out, although from time to time they crash their car to die from the outside in. Leaning on the bonnet, I looked at the valley that opened out beneath my feet and was thinking about the mysteries of life and the fantastical story of evolution, when a crow flew past, cawing, very close to my head.

"Hello, crow," I said telepathically.

But it didn't hear me, or pretended not to.

FIFTEEN

Advantages and disadvantages

On 30 September, Arsuaga had just had a birthday, though he gave every indication of having turned one year younger.

"You're looking well," I said as I buckled up in the passenger seat of his Nissan Juke.

"We should have met at eight," he replied, starting up the car.

It was his way of reproaching me for our having met at nine, owing to my insistence on delaying the appointment.

"For us to meet at eight," I said, "I'd have to get up at five. I'm very slow in the mornings."

"Nothing to be done about it now," he concluded vigorously. "Let's go."

When I saw that we were taking the Colmenar Viejo highway, I asked where we were headed.

"We're going to see some surprising, nearby things," he replied, "because there are always surprising things nearby. Today we're going to tie up some of the loose ends we've left over the course of our meetings. You're going to learn some things that, until now, you've only been pretending you understand."

"I don't pretend, Arsuaga," I said. "What happens is that I understand at the moment you're explaining them to me, but when I get back home I don't remember the logical route by which I managed to reach that understanding."

"That's because you keep on clinging to conventional logic."

"Are we going to be cold?" I asked when I saw a sign for San Agustín del Guadalix, a town close to the mountains.

The palaeontologist glanced over at my outfit and made a doubtful face, from which I inferred that, yes, quite possibly we were going to be cold.

After forty or fifty minutes' driving, we stopped by a roundabout, where it seemed we were to meet somebody. We got out of the car and stamped our feet to warm up. Next to an earth track, there was a sign that said "Cattle Route". Everything else looked like nothing, like sheer nothingness. Though the day was cool, it was sunny and the sky was totally blue, without a single cloud. But that barren combination of a traffic island, a cattle route, and a blue sky proved unsettling.

"This feels like a non-place," I said, receiving no reply whatsoever.

A few minutes went by, and I saw a rabbit making its way unhurriedly across the cattle route.

"Look," I exclaimed, "a rabbit!"

"Write it down," he said, "make a note. A rabbit."

I did make a note. Then I saw that Arsuaga was looking up and pointing at a large bird.

"A vulture," he said. "Make a note of that, too."

Soon, this vulture was followed by another two, and then, at regular intervals, a half-dozen more. It looked like a vulture airline.

At that point, a Kia pulled up next to our Nissan, and from

it emerged José Antonio Vallejo, the forest-keeper who'd been our host in Rascafría.

We greeted one another. A cyclist went by. I saw another rabbit.

"So, shall we go?" I asked.

"We're waiting for someone else," said Arsuaga.

Soon another car appeared (a Toyota, I think), and another forest-keeper got out, who was introduced to me as Gustavo González. After the customary greetings, we all got back into our respective cars and followed forest-keeper González to wherever he was leading us.

Along our route, which was partly sealed and partly dirt road, I kept seeing storks' nests on one side or other, some of them on old chimneys; others on the lampposts, in spite of the defences that had been devised to prevent them lodging there. Then, in the distance, I saw a shadow in the sky, huge and strange, more or less round, completely covering the sun. I understood, as we approached, that it was a flock of birds, the densest I'd ever seen. A huge mass of feathers flapped about over a ziggurat-shaped mountain, which turned out to be the Colmenar Viejo municipal tip.

We stopped the cars and got out to look, in awe, at the spectacle.

"Those silhouettes over there," said forest-keeper González, pointing at one of the birds perched on the edges of the ziggurat, "are the vultures; those other ones, they're the gulls."

"And the ones on the right?" I asked.

"They're the storks that no longer fly to Africa at the end of summer."

Forest-keeper González opened the boot of his Toyota and pulled out a telescope, which he set up on the side of the road so that we could clearly distinguish each group of birds.

"There are two kinds of gull," he said, "laughing gulls and black-backed gulls. And two types of vulture, griffon vultures and black vultures. If you look over there with the telescope, you'll see red kites, crows, and jackdaws."

"If half of these birds decided to attack us," added Arsuaga, "we wouldn't last five minutes."

"Even less than that," replied forest-keeper Vallejo.

I didn't know that there were such things as laughing gulls and black-backed gulls, and I couldn't have told you which were more scary when they were gathered in their hundreds or their thousands, and so far from the sea, which was the mythical space I'd associated them with. Two red kites were either fighting or playing in mid-flight — how was I to tell?

"Could be fledgelings doing acrobatics," forest-keeper González explained.

"Oh, and don't miss that flock of vultures over there," forest-keeper Vallejo pointed out in turn.

There was something endless about it, as we looked this way and that, marvelling that this vast quantity of natural wonders, all of them so beautiful and so powerful, had been reduced to rubbish-heap indigents. Since the ziggurat looked rather like the Tower of Babel, as represented in some classical paintings, it was logical that the range of flying creatures should be grouped together by species because they spoke different languages, the echoes of which mingled in the air like a parliamentary debate where everybody's talking and nobody's listening. They don't understand one another, I thought.

"Pure Anthropocene," said Arsuaga.

We got back into the cars so as to move closer to the dump of Babel and look at it from a different perspective. As we skirted alongside it, I saw a dirty old mattress abandoned in the gutter, then another, and another. How sad they are, mattresses, when they've been written off. The vegetation, which was sparse, grew here and there irregularly like the hair of a patient who has just come through chemotherapy. It was all very gloomy, in spite of the prevailing sun. In the distance, way in the distance, we occasionally glimpsed the Madrid skyline, with its new towers blurred by the haze of pollution. Having left the cars at the foot of a mountain of rubbish, we climbed a ten-metre slope from whose top everything looked even more unsettling, if that was possible.

"The other day," said forest-keeper González, pointing at some nearby high-voltage cables, "we removed a vulture from that power line, it was totally fried."

"What will happen," I asked, "when it's not possible to accumulate any more crap without risking a landslide?"

"The whole thing will get covered in a layer of vegetation," said forest-keeper González, "and they'll have to set up the tip someplace else, which isn't easy. Nobody wants it in their back yard."

"Two thousand years ago," said Arsuaga, "there were lions and hyenas around here."

"Here," added forest-keeper González, "you'll find all the nocturnal birds of prey, and almost all the daytime ones. So you won't see a single rat around."

"I did see a rabbit earlier," I said.

"That's surprising," he replied.

I moved apart from the group for a moment to inspect our surroundings, and discovered a wire fence that protected some train tracks that were half-hidden down a kind of ravine. There were the remains of plastic bags caught on the wire netting, which came from the tip, waving in the wind like flags from some country or other. When I rejoined the group, Arsuaga and the forest-keepers were talking about life. Forest-keeper González was saying, just at that moment, "Each of us has got two lives; the second starts the moment you realise you only have one."

"And then?" asked Arsuaga.

"Then all you want is a house in Asturias."

I felt lucky, as I had just such a house.

"The Zoroastrians," said Arsuaga, pointing at a group of scavengers, "hung dead bodies up in trees for the vultures to eat."

"It's the quickest way of getting back to nature," said forest-keeper González. "I'm a fan of cremation."

"They make diamonds out of the ashes nowadays," added forest-keeper Vallejo.

The vultures started to circle as if we were dead, and a large hairy fly settled on the sleeve of my jacket — a very ugly fly, to tell the truth. I hinted that maybe we ought to leave that place, but they went on talking a little more about life.

"We've gone from shit to philosophy," concluded Arsuaga.

That was when we said goodbye to forest-keeper González, who was kind of like the god of all that fauna. Arsuaga and I got into the Nissan Juke and followed Vallejo's car, since the day, apparently, was not over yet.

"Where are we off to now?" I asked the palaeontologist.

"You'll see," he replied, shifting gear.

"Okay," I said, retrieving my notebook from my pocket, as I sensed he was about to tell me something interesting.

And soon afterwards, never losing sight of Vallejo's car, he did indeed turn to me and say, "You'll be wondering what all of this has got to do with our book."

"Well, yes."

"Okay, so birds have a basal metabolism one and a half times faster than that of mammals the same size. That means a bird that weighs one hundred grams needs more calories than a mammal of the same weight. One and a half times more. Do you remember what basal metabolism is? We talked about it at the Museum of Natural Sciences."

"It's the number of calories," I said, "that an animal needs … when at rest and at a comfortable temperature … to carry out its vital functions and stay alive."

"Right, and what's that got to do with longevity?"

"I don't know," I said.

"Think about it," he insisted. "Something we've already talked about. Remember the example of the mouse and the elephant."

"If the duration of life," I ventured, "depended on the basal metabolic rate, birds should live shorter lives than animals of comparable weight, as flying requires extra effort."

"Their lives should be a third shorter," said Arsuaga, confirming my idea. "It's the theory of kilometres on the clock that we discussed at the scrap yard. The more kilometres a car's done, the sooner it gets scrapped. You wouldn't pay the same for a second-hand car that's done ten thousand kilometres as you would for one with two hundred thousand on the clock."

"And so mice," I added, "whose metabolism is frantically fast, don't live as long as elephants. The more frantically a species lives, the shorter its individuals' lives will be."

"And yet," he said, checking to see that I was writing things down, "all the birds we've seen live shorter lives than mammals of an equivalent weight. They are the exception to the rule."

"At home, when I was little, we had chickens, and I don't remember them having very long lives."

"I'm talking about airborne birds: vultures, crows, kites, condors, albatrosses, eagles. Parrots, to give a well-known example, live a very long time. Flamingos, too; flamingos go on for decades. Gallinaceous birds, which include partridges and pheasants, don't live as long as mammals their size."

"How curious!"

"Isn't it!" said the palaeontologist. "Imagine giving your five-year-old son a hamster as a present, and when he turns forty-five, the hamster's still going round and round on its wheel, with your grandchildren looking on."

"That would be appalling," I said.

"A condor lives for ages, like a hundred years," he said. "And yet it does have a very fast pace of life, much more so than a mammal its size. Flying consumes lots of energy, but they still live longer."

"So there's something about flying that makes birds live longer."

"I don't know," said Arsuaga, "but it is something we need to look into if we're really interested in the subject of old age."

"In any case," it suddenly occurred to me, "birds aren't mammals, after all, are they? Maybe there's some mechanism in their biology that can only apply to them."

"They aren't, but they are warm-blooded vertebrates, like you and me. And along with mammals, they're the only animals that regulate their body temperature by generating warmth inside themselves, so they don't have to depend on it being warm outside, like reptiles and other cold-blooded animals."

"I'm still not sure that the same goes for both us and birds," I said, doubtfully.

"Well, you will be when we get to the place we're going now."

At this point, Arsuaga asked me to close my notebook because he wanted to tell me something he didn't want me to write down.

"That lot up there," he said, pointing at the roof of the Nissan, "don't like it when we make plans that don't involve them. The gods often use heart attacks to punish this kind of insolence. Cancer is more our responsibility, there are lots of cancers that are to do with our environment, but heart attacks are the gods' preferred mode of execution."

"Right," I said, not sure if he was serious or joking.

"Which did we say was the worst of all human sins?"

"Pride," I said quickly.

"Planning something without the gods' permission constitutes an act of pride."

"And where are you going with all this?"

"Yesterday, I saw an interview with you in some newspaper or another, and you said we were writing a book on old age and death."

"We are. Actually, we're finishing it."

"Why do you think people so often caveat what they say with things like 'God willing'? Because they know the gods can't stand it if you don't bear them in mind when you're working on a project. Don't keep telling people we're working on a book until we've finished it — you might jinx it."

"Fucking hell!" I exclaimed. "You're more superstitious than I am."

"I've got proof, is all. Heart attacks are inexplicable.

Suddenly, somebody who didn't smoke or drink, who was perfectly healthy, had just been for a check-up, was living an exemplary life, *bam*, their heart blows up. Why? For making plans without consulting the gods. I know of cases that would make you shudder."

"You probably shouldn't tell me."

"Then stop telling people we're working on a book."

"Okay," I said, with a concern that was not pretend.

In the meantime, we'd arrived at an incredibly beautiful spot, lush with vegetation, belonging to Lozoyuela, a small locality in Madrid's Sierra Norte. If you'd told me I was dead and that this valley was Paradise, I would have believed it, especially compared to the Colmenar tip, from which we had just come and which could easily have passed for Hell. There, in the middle of the countryside, there were another three forest-keepers waiting for us: Beatriz del Hierro, Pilar Moreno, and Jorge Cicuéndez. The palaeontologist pointed at a hill quite far away and said, "That's the Medio Celemín gate — a 'half celemín' being the dry measure you had to pay to pass through it. And that peak," he added, pointing further to the right, "is Mondalindo. See what a great view you get from here of the back of La Cabrera."

One of the forest-keepers told us that we were going to follow the route of the Cañada Real Segoviana.

"To go where?" I asked.

"You'll see," was all Arsuaga said.

So we got back into the cars to follow the new forest-keepers who had come into our lives. The Cañada Real was an earth track reserved for the passage of livestock. Another cattle route, then, but with so much plant beauty around it that it in no way resembled the one we'd seen by the Colmenar tip. The day's bright splendour was reflected mirror-like in the vegetation, injecting our spirits with the sort of optimism you get from certain artificial stimulants, but with none of the side effects.

We soon stopped in a corner of paradise even more welcoming than the previous one. We got out of our cars and contentedly inhaled that clean air, which was also at the perfect temperature, neither cold nor hot.

Then forest-keeper Vallejo, who had taken something out of the boot of his car, handed me a miner's lamp and a helmet. With his help I put them on, fearing the worst.

"What's this for?" I asked.

The answer was to be found a few metres away, since after we'd gone down a nearby bank, which had been hidden from my view, a metal barrier appeared with a door that one of the forest-keepers unlocked. Beyond it, we found some train tracks, and to their right, about twenty metres away, a tunnel of an intestinal darkness that contrasted with the midday light.

"This is the route the train used to go," they told us, "from Madrid to Irún, but it's no longer in use. It started up in 1968, and stopped running in 2000. It didn't have a very long life."

"I took that train to Paris, the sleeper. It was the famous Puerta del Sol," said Arsuaga.

It smelt very pleasant, because down the sides of the old tracks there was a whole host of aromatic plants.

We stepped into the middle of the tracks with our head torches lit up like the eye of Polyphemus, and no sooner had we crossed the threshold between the daylight and the gloom of the tunnel than the temperature dropped at least ten degrees.

The ground, which was very uneven, being made up of bits of granite, required that we place each foot carefully before shifting our body weight so as not to risk twisting an ankle and spraining it.

"These stones," Arsuaga told me, "are known as 'ballast stones', they're used to hold down the sleepers or cement blocks on which the railway tracks are laid."

We were walking roughly in single file. As they got moved by our footwear, the stones made a rather sinister noise, which was multiplied by the hollowness of the tunnel. The walls, of old concrete, I guessed, filtered the water from the mountain we were under, so that the prevailing cold was quickly compounded by a damp that made it even harsher. I kept looking back — I didn't want to lose track of the exit — but the opening of light was getting ever smaller as we ventured further into the bowels of the earth.

"You're sure no trains come this way?" I asked, half joking, half not.

The forest-keeper closest to me smiled while he used his torch to show me the thoracic cavity of a roebuck that must have been devoured right there by some animal.

I felt like I was running short of oxygen, even though the air current, while weak, was constant.

Soon we stopped at a large black stain that appeared on the ground next to the tracks. Apparently it was bat droppings. Forest-keeper Vallejo crouched down and picked up a handful to show them to us. They looked long, like pieces of wool. Then he crumbled them with his fingers and they turned to dust.

Now I understood what we were doing here: we had come to see bats. I said as much to Arsuaga, and he confirmed it.

"Where better," he added, "than in a disused tunnel?"

But the bats would not be seen, however much further we went into the tunnel's darkness. I went on looking back from time to time to check the size of the opening of light at the entrance, which by now was turning, dangerously, into a mere slit.

"Now," said someone, "you can't see them as much because they are currently active, as there are still insects around. To hibernate, they gather in bunches on the roof, and then it's very easy to look at them."

At last, lodged in a fissure in the concrete, we found one, which, alarmed by the crossed beams from our cyclops eyes, abandoned its place of refuge and crossed the space from one wall to another, almost grazing us, like a clot of darkness even blacker than the darkness that filled the tunnel. From that moment, they began to show themselves. Unsettled by our presence, they flapped giddily about in front of our faces.

"They're winged mice," said Arsuaga. "Mammals with wings. Here you have another example of the Anthropocene. Basically, we're inside a neo-cave."

"Last year," said one of the forest-keepers, "we counted over two hundred of them, mostly horseshoe bats."

Since the purpose of our excursion had been to see bats and we had now seen them, somebody suggested we start heading back, a proposal I welcomed enthusiastically, as the point of light that could be made out in the distance was no bigger than the eye of a needle.

Faced with the relief of returning, my nervous tension dropped and I noticed that my ears and hands were frozen. I also felt, on my back, the cold of the constant air current, which I feared would give me pneumonia.

The return walk, perhaps because of my eagerness to get

back out into the light, took longer than the walk in, but as we progressed the opening of light did grow, causing a linear-perspective effect with its vanishing point — which is, incidentally, the perfect term, since, at least for me, what we were doing was trying to disappear from there as quickly as possible.

After thanking the forest-keepers for looking after us and saying our goodbyes, the palaeontologist and I began our drive back to Madrid in the Nissan Juke.

"You'll remember," he said, as we put on our seatbelts, "that airborne birds — so, not the gallinaceous kind — live longer than mammals the same size, despite the fact they consume more calories. But, as well as being warm-blooded vertebrates, bats are mammals. This is what I wanted you to see: a mammal that's airborne and that lives longer than un-winged mammals the same size. Much longer than mice."

"So how long does a bat live?"

"Up to about twenty, despite the fact it has a higher basal metabolism than a mouse. A lab mouse, if it's well looked after, properly fed, and so on — and obviously it has no predators — still won't live to more than four or five. That's a hell of a difference. So here we have a case where we need to find something other than metabolism to explain longevity."

"What, then? The fact that it flies?"

"To answer that," he said, starting the engine, "I'm going to give you a couple of facts about the lives of bats. The first is that they only give birth once or twice a year, and that they take great care looking after their newborn, of which there's usually just the one. The second is that they grow slowly in comparison with rodents or other insectivores of their size."

"All animals that live a long time develop slowly," I said.

"Aha. That's what the naturalist Buffon said in the seventeenth century, and it's what we saw at the Museum of Natural Sciences. All you need to do, for a rough idea of longevity, is multiply by three the time a species takes to reach its full size. In our case, if you multiply twenty-one by three, you end up pretty close to the longevity of our prehistoric forebears. The equation worked out better when Buffon did it, because he still saw men as having to 'fill out' after they reached full height, which can take a number of years more, up to the age of thirty. And I'll remind you, just in case, that longevity is *the maximum duration of a life*, what the oldest member of the tribe lives to. Not to be confused with *life expectancy*, which is the age at which half the members of the tribe die."

"I remember that. But I'm interested in shoring up this idea: about how slow-growing species live longer than fast-growing ones."

"What determines longevity in a species is its rate of mortality," he said, glancing out the corner of his eye to check that I was noting his phrase down.

"That sounds like you're just playing around with words," I said, discouraged. "I don't follow."

"Species with high mortality rates, like rabbits and mice, remember? They grow quickly and reproduce as early as they can, and they produce huge numbers of offspring, because, if they don't, there will be no offspring at all and their genes won't get passed on."

"And how do you ensure low mortality?" I asked, hoping that he might be more concrete.

"Don't get eaten," he said.

"And how do you stop yourself getting eaten?" I insisted.

"By making yourself very big, for example, like an elephant, or like a whale. And to do this, clearly, you need a long period in which to grow. A whale doesn't reach full adult size from one day to the next. Or does it?"

"Of course not. But if you're very big," it suddenly occurred to me, "you need a lot of food."

"Right: if you're small, you don't need much food, but everyone else eats you. If you're big, nobody eats you, but it might be that you can't get enough to eat, and you starve to death. Which is better?"

"Ugh, what a choice ..."

"Another way of stopping yourself from getting eaten," the palaeontologist continued, "is by making yourself very clever by way of your big brain, like with human beings. In that case, you don't need to grow as big as a whale or an elephant. But a big brain also doesn't come about from one day to the next, and its upkeep is expensive, very expensive. That one point five kilos of grey matter you carry around inside your skull is barely two per cent of your overall weight, but it consumes more than twenty per cent of your daily calories — that is, more than a fifth of your energetic budget, in your bodily economy. You can be very clever, but you'll still die of starvation if you don't get the number of calories you need every day to fuel it. A small brain is less of a guzzler, but it provides you with fewer resources when it comes to tackling life. Everything in nature's like this: advantages and disadvantages, compromises, what the English call 'trade-offs'."

"Sounds weird talking about calories in economic terms."

"Ah, it's not quite romantic enough for a temperament like yours," he replied ironically.

"What I'm thinking is, if you acquire the ability to fly, like a bird or a bat, you rid yourself of a whole lot of predators."

"In fact, airborne birds barely have any predators at all. Albatrosses are probably the ones that live the longest, as long as humans. But in order to fly, you need more development time. A bat also isn't made in a day. A bat is a pretty sophisticated bit of aeronautical technology. It uses its hearing to orientate itself in the darkest of nights."

"And if you can live underground," I said next, "you can also escape predators."

"I'm pleased you remember the naked mole-rat we saw at Faunia. Everything we opened up in our first meetings we're going to tie together in these final ones. Oh, and we're still missing one: the giant tortoise from the Galapagos Islands, which we saw at the Museum of Natural Sciences. If you're big and you've got a shell to protect you from predators, you can live for absolutely ages."

"Remind me — briefly, please — why does old age only exist in modern humans, in zoo animals, and in pets?"

"Okay, make a note of this," he said, slowing down, as we'd hit traffic. "I'll use the example of the bats we just saw. Suppose they never got old, suppose they were eternally young."

"Okay, I've supposed it. Go on."

"That doesn't, however, mean they'll never die through external causes. A bat could be in a flying accident, it could fly into some electric cables. It could get into a fight with a fellow bat and end up unable to fly, which would mean dying of hunger in pretty short order. There could be a year with very few insects, meaning it perishes for want of food. As well as accidents, all species are the victims of debilitating parasites."

"We don't give much thought to parasites," I mused.

"Well, they're part of the ecosystem, they're everywhere. Also make a note of viruses, bacteria, fungi, and other external agents that prompt illness."

"Right, I'm following: they don't get old, but sooner or later they die of external causes, like the glasses at the restaurant or Medawar's test tubes."

"Hold on. Now imagine that none of these creatures live to see twenty, all having died due to external causes."

"Okay."

"Right, now — everyone, you and me included, is a mutant. Mutations appear as errors when DNA is being copied. If any mutation brings about developmental disorders, it might then be that we aren't born at all, or that we die before managing to reproduce. But usually these mutations have no consequences, and we live a normal life and pass them on to our offspring, who also live a normal life, et cetera. Are you with me?"

"So far, yes."

"Now imagine a bat is born with a mutation that produces an illness twenty-five years after it's born. What's going to happen?"

"Nothing, because no bats live past twenty."

"That's in nature, almost none live past twenty. But imagine you get a bat, take it home with you, and keep it away from all external causes of death: you remove its parasites, you guarantee it food, you protect it from weather extremes, and so on."

"In that case," I deduced, "it would die at twenty-five, because at that age the mutated gene with deadly effects would be expressed."

"Well, this is exactly what's happening to us modern human beings, as well as to the wild animals we keep in zoos and to our pets: past the age when we'd have died in nature, we start to see the expression of detrimental, belated genes, which have been accumulating over the course of the species' evolution. I'll say it again: natural selection never got sight of these detrimental mutations because they were never expressed, given that their carriers died at an earlier age. Not having got sight of them, there was also no way for it to eliminate them."

"And that's old age."

"That's old age, which is something that never comes up in nature because of the implacable effect of external causes of death."

"Right," I said.

"We'll leave it there for today; I see you're worn out. Enjoy the traffic jam."

"There's only one other thing — it's just occurred to me — about bats."

"What?"

"The fact they hibernate. In other words, they spend several months with their metabolisms down to the minimum, consuming very few calories. Maybe the energy they expend is lower than that of rats, maybe that's why they live longer."

"Don't try to be clever, Millás. It turns out that bats, in Mediterranean Spain, are actually hibernating *less*, because climate change means they have access to insects almost all year round. Plus, there's a variety of tropical bat, a large kind called the 'flying fox', that lives off fruits and doesn't hibernate at all because you don't get the same seasons in the tropics. Curiously, they're the ones that live the longest, and the ones whose development is the slowest, on account of their size. I tried to show you some at Faunia on our first outing, but your inner lunch-bell rang and you stopped listening to anything apart from your hunger."

Hunger, I said to myself, unable to stop thinking about the materialistic idea of the energy budget.

"I'm going to drop you at Plaza de Castilla," said Arsuaga. "I'm late for a meeting."

SIXTEEN

There's nothing pre-programmed here

The place was so beautiful that I felt a little ashamed asking the palaeontologist what we were doing there, as if you could only go to beautiful places in order to do things. The place in question being a vast meadow with hills around it on all sides, creating the effect of a large pot with not very high walls, and us at the bottom. "Us" being Arsuaga; Ángel Gómez, the director of the Cabañeros National Park; and me.

Evening was falling over the grassland and over the range of hills around, but it hadn't yet started to get dark. A silence was following us, the way we're followed by our shadows. We couldn't detach ourselves from it, not even by speaking, since our words just got absorbed into its matter, whatever the matter of silence is.

"Right here, directly below this exact spot on the other side of the earth, is the Tongariro National Park, in New Zealand," said the palaeontologist.

"Tongariro?" I repeated, but adding a questioning tone.

"A sacred place for the Māori," Arsuaga explained.

We'd left the 4×4 at a strategic point in the centre of the pot, from which there was a view of the landscape that could have awoken religious feelings in even the least metaphysically inclined of people. I mean religion in its etymological sense, which comes from the Latin verb *religāre* — that is, relating to the bond with one's surroundings, especially with nature, of which we humans are no longer a part. In the middle of that seeming African savannah, a nostalgia arose for a distant past, a time when we did still constitute part of what we were now gazing upon, our awe like that of an amputated limb as it looked back, many years after the separation, on the body from which it had come.

Ángel Gómez said, "This landscape is known as the 'Spanish Serengeti', owing to its resemblance to Tanzania's National Park."

"So it's not foolish to imagine ourselves in Africa?" I asked.

"Oh, no," he said. "Actually, if you were brought here with your eyes closed, you'd feel, when you opened them, like you'd been dropped into the middle of the African savannah."

Amid the thick, tall grass, so dry from the lack of rain, we could see that the landscape was dotted here and there with the antlers of the odd deer whose fallen body was tangled up in all the vegetation. But it also wasn't unusual to see roving clusters of live deer, or smaller groups, in threes, usually a doe with two fawns born the previous spring.

The Cabañeros National Park, with a surface area of forty thousand hectares, is a protected natural area located in the provinces of Ciudad Real and Toledo, in the autonomous community of Castile-La Mancha.

I didn't need to ask why we'd come here, because Arsuaga told me right away.

"I've brought you here so you can witness one of the most magnificent manifestations of deer when they're in rut."

"The rutting calls?" I asked.

"The rutting calls," Arsuaga replied. "It's also called 'roaring' — the sound the stags make to get the does' attention."

"I don't hear anything," I said.

"The party doesn't start until dusk."

And so while we waited for dusk to come, we wandered the vast grasslands. The deer, as we passed, stopped and stared in a state of high alert. They looked at us, we looked at them, and you got the sense that in that crossing of stares, the spark of meaning was always just about to catch.

"The females," Gómez told us, "the does, come on heat once a year, now, in October. Which coincides with the calls, of course."

Sprinkled around the savannah floor we could see countless little quartzite pebbles, which came, Arsuaga explained, from the erosion of the Montes de Toledo, the hills that made up the range above and around us.

"'Raña', that's what we call this kind of terrain. Not a good place for rabbits, because it's so difficult to dig their warrens."

Here and there, it was possible to make out, scattered around the plain, groups of holm oaks, some of which, due to the growing darkness, reminded me of human figures that were misshapen or in a state of decrepitude.

"We don't give the deer any food or water," said Gómez. "They get by on their own, just as in the wild."

I did think, albeit without saying so, that the syntagm "protected natural area" was a slight contradiction in terms.

"But they don't have many predators," said Arsuaga, as though he'd been reading, as he often did, my mind.

"Do they self-regulate," I asked, "so that they don't give birth to more than they have space for?"

"You must be kidding!" said Gómez. "We've got to catch about fourteen hundred a year. Some get sold to other game reserves; the rest, the ones that are left, go for human consumption. They reproduce at a rate of knots. Currently, there's one deer for every four hectares, when the ideal would be to have one for every twenty."

"That self-regulation thing," added Arsuaga, "do try to get that out of your head. Nature doesn't self-regulate. Why would it need to, if there's no purpose to any of it?"

"For the good of the species?" I asked, uncertainly.

"I just can't get these romantic ideas out of his head," said the palaeontologist to Ángel Gómez. Then, turning to me again: "Natural selection couldn't care less about the good of the species, I've told you a hundred times. What's in play here is, firstly, the survival of the individual at all costs. I'll say it again: *the survival of the individual at all costs*. And once it's survived: self-perpetuation. Everything else — the ecosystem, the population, the species — simply isn't natural selection's concern. If human beings didn't play the role of predator in this national park, the ecosystem would be totally overrun by deer within days. That's right: fourteen hundred fewer deer annually and there's still way too many."

"They did an experiment on a Canadian island with wapiti deer — elk," said Gómez. "They left them living there without any predators, and the deer began to multiply and multiply, until they'd eaten everything there was to eat, so in the end they became extinct."

"Write that down," said Arsuaga.

"I'm on it," I assured him. And while I was writing it down,

I thought about the jungle of neoliberal capitalism, where the "invisible hand of the market", far from regulating things, allowed the ecosystem to deteriorate to the extremes we were seeing now.

"What's the proportion of stags to does?" the palaeontologist asked .

"The same number of stags and does are born," replied Gómez.

"What an economic waste!" exclaimed Arsuaga, to me.

"Why economic waste?" I asked.

"On our way here," he said, stopping walking so that I'd concentrate on what he was saying, "we saw various livestock farms where they'd have had something like one bull to every thirty cows. Logical, right?"

"Totally," I agreed.

"Why?"

"You explained it to me yourself. Why have a bull for every cow if one bull can cover thirty, or forty, or however many?"

"So, why do you get the same number of males being born here as females, if a single bull can cover thirty or forty cows?"

"I think," I ventured, more as a suggestion than a definite assertion, "that it makes for competition among the males, so only the fittest get to mate. And then the best genes get perpetuated."

"In other words, because nature's so very wise?" Arsuaga queried ironically.

"Kind of," I said.

"But nature isn't wise."

"It has no purpose, no direction, no aim, and it doesn't self-regulate," I added, as if reciting a litany.

"Okay, let's try just getting you to repeat it, and see if that way you end up believing it."

"But then why are there the same number of males as females?" I asked. "Why this anti-economic surplus of males?"

"Imagine a deer colony," said Arsuaga, "in which one male is born to every fifty females. Then a female appears with a mutation that means she gives birth to only males. Since there was previously a deficit of males, this female who gives birth to only males has her progeny guaranteed, which is what this is all about, securing the continuity of the genes. The males, in this situation, are increasingly abundant, because it's better to be male than female. If you have male offspring, you'll have *lots* of offspring, and they in turn will multiply ... Are you with me?"

"I am."

"It's another thing if we're talking about a monogamous species, but we aren't here. Okay?"

"Okay."

"So then the proportion of males goes on increasing until a moment comes when there are more males than females. In this case, if you were a female, it'd be in your interest to produce females because they, unlike the males, are guaranteed to be covered, and therefore to produce more offspring. Then say a female appears with a mutation that means she gives birth to only females, et cetera. And in this way, out of the struggle between individuals over who's going to have the most grandchildren, you get this balance that finally means the same number of males and females being born. Remember, Millás, this is about the generation *after* next, because there's no point in having children if they don't reproduce."

"So it isn't that nature's wise," I concluded.

"Of course not, it's purely about mathematics, about deficits coming into balance. If nature were wise, I'll say it again, in this polygamous species you'd get one male being born for every thirty or forty females."

"You invest in males when there a lot of females, and in females when there are a lot of males," I said.

"That's it. And in this way, mathematically, it arrives at one to one. All evolutionary biologists agree that there's no thought in nature, no wisdom, nor any other such nonsense. Pure genetics and pure calculation of probabilities, nothing more. Full stop."

While we talked, the sun had been declining lazily, reluctantly. Now it seemed to be that unsettling hour when the day had stopped being day, but this didn't mean it had become night. The east looked like it wanted to grow dark but was being denied by the now almost horizontal rays from the sun, which was still there suspended in the west. We saw a stag approaching three does.

"That one, to go by the number of points on its antlers," said Gómez, "is second fiddle. The alphas are exhausted from such a lot of screwing, from such a lot of fighting, and they've withdrawn. The females, on the other hand, have eaten acorns, they've recovered and are contemplating the arrival of these second fiddles, though they're certainly already pregnant by the alphas. The solitary males don't usually get them. They approach, see another male that's more vigorous, stronger, and they retreat."

"Poor things," I said.

On the ground I saw a huge set of antlers, which I assumed came from a dead stag because I thought — incorrectly — that antlers lasted a whole lifetime and that the number of points corresponded to the animal's age. But apparently not: they shed them once a year and grow them back like a stalagmite sprouting from inside their skulls.

"How long does it take them to grow back?" I asked.

"A couple of months," said Gómez. "Which takes a really serious expenditure of energy."

"And what does the number of points depend on?"

"Mostly on how they feed. When ours shed theirs, we leave them where they've fallen because the younger deer suck on them for the calcium. But we've got to watch out for poachers, who steal them to sell to the Chinese. In China, stag antlers are used as aphrodisiacs."

I looked at my watch and it was 8.30. The disc of the sun was no longer visible, but amid the prevailing darkness there was still a bright glow, like fresh blood, behind the mountains.

Then we heard the first howl.

This first one was followed by another and another. I interpreted them as guttural sounds of suffering, as something between an entreaty and a demonstration of power. The concert grew, while the darkness intensified and the first stars appeared in the sky. You only had to avoid looking up for three minutes and the next time you raised your eyes another half-dozen would have appeared, as if somebody were placing them up there, gradually, as night progressed. Soon the bloody glow in the west disappeared altogether, and we three humans were left in total darkness, which was very dense owing to the lack of light pollution. Instinctively we moved closer to the 4×4, as if it were a defensive weapon, and as we leaned on its bodywork we let ourselves be shaken by the bellows that came from every corner of the park, passing through the air and crossing one another above our heads to form an ancestral web of venereal expression.

Amid that total darkness — of which we, like three

shadowy masses gathered around the car, were ourselves now a part — the stags would already be clashing antlers, sometimes with fatal results, in a fierce battle for the survival of their genes.

"Look," said Arsuaga, raising his dark arm towards the sky, "Jupiter and Saturn are clear as anything."

Not long afterwards, Ángel Gómez drove us to a kind of former farmstead that was right inside the park itself and had been restored to accommodate occasional visitors like Arsuaga and me. Once we were alone, with almost teenage excitement we went around the house, which alternated large living rooms, which in their day had been stables, and impenetrable little corners. There was a convent-like austerity to the bedrooms and the bathroom, and likewise the kitchen, on whose counter we'd been left food for our dinner: a large loaf of white bread, a venison salami and chorizo, a sheep's-milk cheese, and a jar of marinated partridge, along with a bottle of cabernet sauvignon that we gladly uncorked to accompany the food.

After dinner, we stepped out into the mysterious, dark countryside, both of us a touch overwhelmed by this strange intimacy we'd rushed into. Since he's not much inclined to outpourings of a personal nature, the palaeontologist saved the situation by asking me to look at the sky so he could show me the constellation of Orion, the hunter, with his belt and his bow. Once we'd found it, he said, "Can you see that red star, the very bright one, on the hunter's shoulder?"

"Yes," I said.

"That's called Betelgeuse. It's a red giant that's going to explode at some point and turn into a supernova. Because of its enormous size and its proximity to earth — it's only six hundred and fifty light years away — it's going to be the biggest supernova humankind has ever seen. It'll be visible even in the daytime."

"And how long till it explodes?" I asked.

"It could happen any time. It already might have, or it might be millions of years before it does. There have been other, smaller supernovas before. The most famous one happened in the summer of 1054; there are records from Chinese astronomers about it."

"Right," I said, not taking my eyes off Orion, the hunter from Greek mythology who claimed he could wipe every animal off the face of the earth.

I remembered then that my father, on summer nights, had insisted on showing me the constellations and that I would look up and pretend to see them, though I never managed to, not once, not until that night at Cabañeros, under the tutelage of the palaeontologist.

We spent another hour in the open, finishing off what was left of the cab sav, shielding ourselves from that unexpected intimacy and alert to the sounds coming from the hermetic darkness, forty thousand hectares of it, where we found ourselves, as if lost.

"A wild boar?" I'd ask. "A fox?"

At 7.30 the next morning, Juan Antonio Fernández came to fetch us. He was the senior environmental officer (what would once have been a "head keeper") and patrol coordinator.

It was still dark at that hour, so we got into his 4×4 and went back again to hear the resonant manifestations of those amorous courtships that were coming from every corner of the park and which were strangely synchronised so that, after a while, instead of hearing one bellow here and another there, what you'd hear was a chorus of entreaties disguised as vocal potency. I thought the calls had perhaps been misinterpreted: maybe they weren't a demonstration of sexual vigour, but a show of helplessness designed to arouse pity in the females.

Of course, I wasn't going to mention such a romantic impression to the palaeontologist.

In the dark place where we'd stopped the car, it was impossible to make out anything beyond a couple of feet away. If I stretched out my arm, I could no longer see my hand: it was in view, yet I couldn't see it. The darkness was, in any case, extremely dense, like a block of solid matter that the harsh bellows from the animals seemed to perforate, filling it with tunnels. It was striking that while we were there motionless, and still numb from the early-morning cold, all around us was the pure ardour of copulation, pure animal desire, pure impulse, pure life.

After eight o'clock, as it grew lighter, silence gradually returned to the land. I looked eastwards and saw the outline of the Montes de Toledo against the red-white glow produced by the sun that was still hiding behind them. Then we dared to venture a little further from the 4×4, and moved about the environs of that domestic Serengeti, its grass damp with the dew.

Juan Antonio Fernández, the head keeper, to use the old term, came over to show me a video he'd recorded on his phone a few days earlier on which you could see two stags whose antlers had become completely interlocked. One of the stags was dead, but they'd got so tangled up in their test of strength that the living one, despite his desperate efforts, couldn't get free. While he shook his head desperately left and right, you could see a couple of wild boar, having managed to get at the dead one's guts, devouring it before the living one's eyes.

Juan Antonio explained to us that the forest-keepers were planting a lot of holm oaks, which are the native tree.

"If you tread on the grass," he added, "you'll see that, though at first sight it seems dry because it hasn't rained much, underneath you can already detect the green, thanks to this morning dampness we get here."

Around 8.15 a.m., there was plenty of brightness, though the sun had not yet appeared over the hills. Ten minutes later it showed itself, a yellow disc that would blind you if you turned your eyes to the east. Shortly before 9.00, its rays made the whole pasture shine like a mirror, casting very long shadows where it intersected with the bodies of the holm oaks. It felt so good exposing oneself to those rays to shake off the mid-October cold, with autumn already embedded in our hearts.

"In winter, temperatures can drop to minus four," said the head keeper. "Then the deer vanish up into the hills and live off rockrose, doing a lot of grazing. They come back down in the spring, but they don't come back weakened, because on the whole they've fed well."

We saw a group of eight or ten lion vultures all flying in the same direction.

"Maybe there's a corpse somewhere close by," said Juan Antonio. "But we're going to see the black vultures."

"This," said Arsuaga, taking in the landscape, "is the most Palaeolithic thing it's possible to see."

Back in the car, we passed close to a stag that was obsessively beating its antlers against the trunk of a holm oak to make the acorns fall out. A short while later, we reached the side of one of the mountains and the head keeper set up a telescope so we could see an eagle whose silhouette was outlined, with great authority, atop a summit. After looking at it, awed by its power, we considered the vegetation around us. There was rockrose, there was myrtle (very fragrant) and thyme and wild olive and cork oak and arbutus and narrow-leaved mock privet and heather. The head keeper seemed to be creating all that thick vegetation as he named it.

We also discovered a very reedy pool that had been filled from the dammed-up waters of a stream.

"Here," said Juan Antonio, "you can see frogs, otters, and a lot of small fish that only occur in very pure, very healthy environments."

It smelled unimprovably good, thanks, among other herbs, to the marjoram. The lichen, which were so abundant and so various, made veritable gardens.

We climbed the hill on foot in search of the black vultures. Every once in a while, we stopped to eat arbutus berries, which were perfectly ripe.

"We'll get up to nine hundred metres," said Juan Antonio. "There's an impressive view of the pasture from there."

We also tried some acorns, but they were still green and bitter-tasting.

At 11.00, the sun was already halfway up, and it was starting to get hot. We took off our jackets and jumpers, remarking on the huge differences in the daytime and night-time temperatures. Then, right in front of us, a pair of black vultures appeared, performing a flying display that looked rehearsed.

When we stopped to rest, the palaeontologist enlightened me: "The bulk of an ecosystem is the vegetation, which is at the base of the ecological pyramid and forms the largest part of the overall biomass. Next level up, you get the primary consumers, who feed on this vegetation: deer, boar, and all the other herbivores. After that, you get the wolves and the eagles, the ones who eat the primary consumers, and who we call, logically enough, secondary consumers ..."

"There are fewer of these all the time," said Juan Antonio. "In Cabañeros, we have three pairs of imperial eagles. Very few lynxes, because there aren't any rabbits, but there are lynxes in nearby estates where the ground's softer and the rabbits can make their warrens."

"Carrying on up the pyramid," Arsuaga went on, "the secondary consumers, the predators, are followed by the super-predators — lions, for example, when there still were lions, back in the Pleistocene. At the apex, you get the necrophagous creatures, represented here by the vultures, which never go a day without a meal. Now, the higher up in the pyramid a predator is, the more territory it requires. Imperial eagles need vast amounts of land. The pyramid remains more or less stable so long as no disasters take place. In the Ngorongoro Conservation Area in Tanzania, for example, which is about the same size as Cabañeros, how many lions do you think there were in the first census they did?"

"No idea," I said.

"Twenty adults, plus their cubs and the juveniles," said Arsuaga. "That was eighty years ago. And how many adults are there now? Twenty. There aren't any more because the territory doesn't provide for any more. The numbers might fluctuate in a bad year, but the point is, population sizes tend to stay as they are. It's what ecologists call 'ecosystem capacity'. There's only so much room for each species."

"Right," I said, trying to catch my breath without anyone noticing.

"We call this phenomenon 'demographic equilibrium', although that's utopian: *demographic equilibrium* doesn't really exist. There are times of increased abundance and times of scarcity, so we talk about an average. In a good year, there are more deer; in a bad year, fewer. If a population is demographically stable — and Spain's, for example, isn't — that means every couple having two offspring. The human birth rate in Spain is less than two, so the population will decline if there's no immigration. Parents are dying and leaving one and a half children. Whereas, when the birth rate exceeds three descendants per couple, that's when you get a baby boom."

"In order to be demographically stable," I concluded, hoping to show that I was making progress, "you've got to have, on average, two descendants per couple, generation after generation."

"And in fact," he said, "on average, not everyone does have two, because of this thing called natural selection. At best, people have eight, and at worst, none at all. But in the end, in a stable population, two offspring per couple always survive."

"And what happens to the rest?"

"I've just told you: the rest get eaten by predators or die from starvation, or thirst, or illness, or in accidents, or because they freeze to death, whatever. This is what natural selection consists of: lots of individuals being born, but only two surviving and reproducing. There's plenty of ways to die. Look what happened to those two stags who got their antlers locked together. What a stupid way to die. But then again, is there really any 'clever' way?"

"There are two left, always, mathematically?" I asked. "Seems pretty unlikely."

"If the population remains stable, then yes. In the sixteen

thousand hectares in this park that are managed by the state (the rest is private estates), there are four thousand deer, half male, half female. Out of the females, a little over fifty per cent are reproducers because the others are still too young. So, let's say, fourteen hundred fertile uteruses. If every uterus produces one descendent a year, and no deer die, you end up with a surplus of fourteen hundred a year."

"Which is why they've got to take them away," I said, "since they don't have any predators and so on. It means this natural space is also kind of artificial."

"The important part is that, by hook or by crook, two must always remain. Write that down."

"Doesn't that contradict the idea that spaces don't self-regulate, which is what we were saying yesterday?"

"Of course they self-regulate."

"Yesterday we were saying they didn't."

"No, it's *species* that don't self-regulate. The deer don't regulate themselves. It's the wolves, or the people managing the park, that regulate them. But call it what you want, the fact is that a structure's established, an ecological pyramid. The population numbers stabilise, and the ecological pyramid is maintained. At the bottom of the pyramid, there's the vegetal mass, which weighs however much; next floor up, the plant-eaters, that weigh however much; next floor, the carnivores, that weigh however much; and on it goes, all the way up to the apex. In a place that's demographically stable — let me just emphasise that: *demographically stable* — every couple in every species, however you want to put it, produces the equivalent of a couple that can reproduce. They have many more offspring, but lots of them die in accidents, fights, from starvation or cold, they get eaten ... The end result is always the same: one reproductive couple for every reproductive couple. All the species in the ecosystem arrive at the same outcome, just by different routes."

"Human beings," I said, "break those rules."

"Because human beings invented agriculture and farming. If you were a farmer and had a few fields in Asturias, you'd know how many cows can be maintained with the grass those fields produced. It's a different thing if you give them feed as well. But let's look at some other examples: mice, for instance, which are food for just about everything else. They can live to three years old. They mature quickly and have lots of offspring, and hope to have reproductive descendants during those three years. Deer live to twenty and do the same thing, because their mortality rates are lower. Why are their mortality rates lower? Because they're bigger, they've got antlers, and they're able to run at incredible speeds — even faster than horses — all of which they manage to do due to a longer maturation process, and because the mother gives the offspring a lot of care, feeding them and protecting them. That's why they can only have one a year. How does our story end?"

"How?"

"A mouse lives for three years and produces lots of offspring, out of which, in the end, what with one thing and another, only two survive. A deer lives for twenty, and in the end only leaves behind a single reproductive couple. The result, on a geological scale, over time, is identical. Look, Millás, I've spent a whole year telling you the same thing in different ways, but I still don't know if you're getting it."

"I do sometimes," I said, abashed.

"Picture the following conversation between a rabbit and an elephant during the Pleistocene ...

Rabbit: My, Elephant, how long you live!

Elephant: That's because I'm clever and I grew big and I don't have very many predators.

Rabbit: And is that what matters?

Elephant: To me, yes.

Rabbit: But if you die without any offspring, your genes won't continue, and it will be as though you never existed. Is that really what you want?

Elephant: Not at all, it's about continuing the line.

Rabbit: Which means you aren't so clever after all, Elephant, because it took you seventy years to produce two reproductive offspring, which is the exact same number I manage in three years. Plus, there are far more rabbits than elephants. We outnumber you.

"At this point," continued Arsuaga, "some humans come along, and the rabbit talks to them.

Rabbit: My, you humans are very clever. You paint such beautiful animals on the cave walls.

Humans: True, true. We're very wise.

Rabbit: How many reproductive offspring do humans have, on average?

Humans: Two.

Rabbit: Same as me, but it takes you sixty or seventy years. It only takes me three.

Humans: But we live for longer than you.

Rabbit: But life isn't about particular individuals living longer, it's about the survival of their genes. Maybe you aren't so clever after all, because there are more of us than there are of you. This is rabbit country, not human country.

"So which out of the three is the cleverest?" asked the palaeontologist.

"Which?" I asked back.

"*Homo sapiens* emerged three hundred thousand years ago. European rabbits have been around for half a million years. If we're still around in, say, ten thousand years' time, and elephants are extinct and we're on the point of being so, just as lots of large, long-lived species have become extinct before us, we'll still have no option but to say that the rabbit's the cleverest of all, because it's still going strong."

"Different strategies," I concluded, "for obtaining the same results: continuity of the genes."

"Different strategies," said Arsuaga, "for maintaining the germ line, of which every individual is the outer shell. And write this next bit down, because it's very important."

"Ready."

"According to the Darwinians, there's no reason for old age. Your bones are made to last for seventy years."

"That sounds like pre-programming to me."

"It sounds like it, but it isn't. The thing is, humans in the Palaeolithic didn't live to beyond seventy. When you and I visited the scrap yard, I used cars as an example. If due to extrinsic causes — because we drive them into the ground — no car lasts for more than ten years, what sense would there be in producing parts that last for twenty? It all *seems* like pre-programming, but there is no pre-programming: there's natural selection. In science, Millás, you always have to think the other way around. It may seem like the sun revolves around us, but it doesn't. It may seem like the earth is flat, but that isn't true either. It took us centuries to realise these two things. Remember what Niels Bohr said to one of his followers: 'It's a crazy theory you've come up with there, but not crazy enough to be true.'"

"Okay," I said.

"In natural selection, always bear in mind, the only thing

that counts is the individual. There's nothing that goes against the individual or its genes."

"Natural selection never works for the species?" I asked.

"Rather it's that the species doesn't know what it is, it's merely an abstraction. Natural selection gives rise to the best individuals — by 'best' we obviously mean the ones best adapted to their corresponding ecological niches. That is, the best rabbits, the best lynxes, the best deer, the best elephants. The best within their situation. And so which ones are going to have the most offspring?"

"The best ones," I said, "which will result in the best adaptation of the species."

"As a collateral knock-on, yes," replied Arsuaga, "but not due to any pre-programming. There's no thinking mind here. Natural selection in fact doesn't select: it just sieves things out. Write that distinction down if you think it makes it clearer."

"It's diabolical," I said. "To get back to our subject: *everything* seems to suggest that old age and death are pre-programmed. You can't tell me they're not."

"It's not about how everything *seems*. It might look like the earth's flat — but that perception is false. So, to be quite clear, we must get back to the external causes of death: a rabbit won't live to more than three as long as there are predators around and limited resources. There isn't enough food in the ecosystem for all the rabbits that get produced. Therefore, whatever happens *after* those three years makes no difference: natural selection never gets sight of it, and does nothing about it. From that age onwards, from the age when they're already supposed to be dead, the mutational burden we've already talked about begins to manifest: arthritis, cataracts, all that."

"And that's old age," I said.

"Exactly."

"So old age doesn't have any general cause?"

"No. It's a case of different genes that come into play here and there, provoking deficits that, in our case, modern medicine is making up for."

"Partially."

"Partially, yes, but it's doing okay. You get a hip replacement, you get a new intraocular lens, you get pills for cholesterol and blood pressure, a cardiac stent, a pacemaker ... All these different sticking plasters mean we live longer and better lives. But at a cellular level, nothing. A lot of value's placed on the fact that many people born today will be able to live to beyond a hundred. That's why I always say I'd love it if death *were* pre-programmed, because, if so, all we'd have to do is figure out how to de-program it."

The view of the pasture, from the top of the hills, was indeed awesome, even seen through a couple of worn-out old lenses like mine.

When the black vultures decided to bring their aerial display to an end, we began our descent.

SEVENTEEN

The Red Queen

Ever since the palaeontologist and I first took on this project, I've been looking out the window every day to watch my old age approaching. I did know that it actually came from inside, but the body is full of internal courtyards that the windows of modern clinical medicine allow us to peer into: tests for analysing blood and urine (and even faeces), CAT scans, X-rays, ultrasounds, MRIs, EKGs, manual examinations ... I submitted to all of these in order to ascertain the health of those internal courtyards, whose condition, despite the odd damp patch and some obvious chipping, wasn't all that bad.

I worried that old age might surprise me away from home, on one of my frequent work trips, the way I suppose a young woman might fear her first period catching her by surprise in school sports, in full view of everyone. I started to be a little afraid of airports and train stations, in case old age hunted me down in one of those non-places, far from the protection of my family and the shelter of those fetish objects that surround me while I'm writing.

Old age, by Arsuaga's calculations, should have caught up with me already, but it wasn't here yet, or at least I hadn't noticed it in myself the way I'd seen it in most other people of my age. It's true that travelling tired me out, even exhausted me, but it's also true that after a few hours' sleep I was back to feeling ready to dive into some new project, some new physical or intellectual adventure. I felt, in short, like that soldier in *The Tartar Steppe*, the Dino Buzzati novel in which an officer has to lead the defence of a fort that, according to his superiors, is just about to be attacked by the enemy. Years then go by without the attack occurring, to the surprise of the officer, who's put his life in the service of something that does not happen in reality, but which keeps happening in his mind. My old age was not happening in real life, but was a constant presence in my imagination.

Days before launching into the adventure of this final chapter of our book, SER radio invited me to do a report on a cruise ship, which meant I had to fly to Rome, where I would board the *Costa Firenze* to travel Rome–Naples–Majorca, after which I would fly back to Madrid. I accepted: I never say no to an offer of work, for fear I might be saying no to something else hidden beneath that offer.

Nothing is what it seems.

The night before my departure I barely slept for fear that old age might suddenly show itself and I'd have to call off the trip at the last minute. I got up at 5.00 a.m., since my Rome flight was leaving early, and peered briefly into my internal courtyards, which were all in order. At Terminal 4 of Madrid–Barajas Airport, I watched myself, suitcase in hand, walking down its mile-long corridors with the agility of a young man. I didn't just feel well: I felt euphoric. My travelling companion, Francisca Ramos, with whom I usually do these kinds of radio reports, even drew my attention to it.

"Well, you've woken up positively radiant today."

I was about to tell her I hadn't slept, but I kept quiet lest my exceptional mood might prove a mere passing excitement. Sort of like the improvement that dying people experience on the eve of their deaths.

But the important part of this story happened on the boat, where I had the opportunity to experience life in captivity, about which the palaeontologist had told me so much when discussing zoo animals, which not only are free from the external perils found in nature but are also watched over by vets and keepers generally monitoring their health. As a result, they live longer than they would in the wild, hence their suffering chronic ailments that natural selection would have avoided if they'd met their deaths when it was their proper time.

The daily life of the individuals of the human species constitutes one of the known forms of captivity, since, as has been explained, we belong to a self-domesticated kingdom. But we aren't aware of it till we're afforded a more intense experience of captivity, like the one I went through aboard the *Costa Firenze*. It was not a prison, of course. I must clarify the difference between captivity and imprisonment. By "captivity" I mean that of animals living in protected natural spaces, napping in modern zoos that imitate the habitats they come from — animals including our domestic animals: cats, dogs, hamsters, birds …

Years ago, I had a canary, and one day I opened the door of its cage. After a little while, the creature came out and flapped, terrified, around the living room until it found a way to get back into the cage, pulling the door shut with its beak. As it had been born in captivity, I deduced that the cage was like a part of its body and that, finding itself outside it, outside its body, it had a panic attack.

The memory of the canary reminded me of the visit the palaeontologist and I had paid to the Cabañeros Natural Park. Even then, I'd wondered how far a so-called "protected" park could also be described by the qualifier "natural". In reality, it was only natural in part, since they needed to extract fourteen hundred deer per annum to keep it in balance. In spite of these deaths, which replaced those that would be caused in a truly natural environment by accidents, infections, or predators, we could say that the deer lived well, if perhaps in a constant state of surprise at never having to flee from any carnivores.

That was what happened to me on the cruise ship, which was why I decided not to go ashore when we stopped in Naples (there was a planned excursion): I wanted to experience to the fullest the captivity I've been talking about for the last few lines. I should say quickly: I felt like a hamster in a very *good* cage, a cage with every convenience to which a hamster born in captivity could aspire. Each day, by circling the perimeter of the giant ship, with the same tenacity with which a hamster exercises in its wheel, I walked the ten thousand steps my doctor had prescribed me. The morning breeze and the sea view heightened my senses. There were other hamsters like me doing the same exercise. Some of them would stop to use some of the many pieces of exercise equipment distributed around the deck of the boat. There were also those who played minigolf or did stretches on devices designed for that purpose.

One day, after my exercise on the wheel (on the deck), I went for a haircut. The young stylist "scolded" me for the wrinkles under my eyes. But it was an affectionate sort of scolding, like we give the cat when he's misbehaving. What I mean is that I didn't take offence; on the contrary, it felt like a show of kind concern.

"Do you use anything?" she asked.

"A cream," I lied.

She went off and came back soon afterwards with a small bottle, a drop of whose contents she applied to my wrinkles. Then she spread the liquid around, and pattered on my skin with the tips of her fingers to help the elixir really get inside. Then she asked me to look at myself in the mirror and — a miracle! — my wrinkles had disappeared. I found myself compelled to purchase the bottle for sixty-one euros, which pained me, I suppose, the same way it pains a pet dog to give his paw to his master whenever he's asked (each of us pays as he can for whatever tidbit his master offers him). That is to say, I did it with pleasure because I was starting to enjoy the delights of life in exaggerated captivity (for the haircut, I paid thirty-five euros, which is kind of like raising both paws at the same time).

I ate very well at one of the countless restaurants on the ship, and while I was having my meal alone, with a hamster's concentration, I realised that everything there was a kind of "as if": as if the frescos on the walls of the spa (where I also sought relaxation) were Roman, as if the golden taps were real gold, as if the plastic wine glasses were crystal, as if the fake marble columns were real marble … I recalled those terrapin pools that have a fake palm tree in the middle. I was in a turtle pool for humans, where the palm trees were also plastic, but where everything, if you surrendered to it, evoked the life of luxury that we're sold by travel agents' posters and consumer society generally. The deer in the Cabañeros Natural Park also spent their days "as if" they were free, just like the lions in the zoo, with their daily diet regulated by the vet.

I couldn't die on the cruise. Animals in captivity don't die, because they always get replaced.

———

Within a few days of my experience of life in exaggerated captivity, Arsuaga asks to meet at 7.45 a.m. outside his house. He doesn't say where we were going, but advises me to wrap up warm.

When we're inside his Nissan Juke, he tells me that the car has neurological problems, though it's still working perfectly well mechanically.

"I just have to supply the neurons myself," he adds, to reassure me.

As we leave Madrid on the Burgos highway, the conversation winds this way and that, with nothing really catching my attention, until he tells me that in Norway, since it has stopped snowing in the winter, the number of suicides has gone up.

"How so?" I ask.

"The snow is so luminous," he says, "it lifts the spirits. Without any snow in those latitudes, it's just dark, dark, dark."

I write this down because it feels like a piece of information worthy of a dystopian tale about climate change.

"My daughter lives in London," he adds, "which isn't Norway, but it does get dark at four in the afternoon, so she's bought a screen that imitates sunlight. They're all the rage. You have to put them up in a strategic spot in your home as a way of dispelling the gloom."

The light screen, I think, is an "as if". "As if it were sunny", to be precise.

At the Somosierra pass, the temperature is two degrees, and a threatening ceiling of low clouds hovers above us. As we begin our descent, we run into a cluster of thick cloud that the Nissan's headlamps struggle to pierce. We move slowly, focusing on the rear lights of the car in front.

"Shame we can't see the vegetation for the fog — the autumn colours here are second to none," says Arsuaga. "Forests like this one are called 'butaneers' because when butane gas came in, people stopped felling them for firewood."

The stress of our driving blind through that gloomy fog ends up plunging us into a doomy silence. *Doomy* and *gloomy*, I think, sound too alike to be living so close together. Mentally I seek out a synonym for *doomy*, and what occurs to me is *ominous*, which always makes me laugh, I don't know why. But anyway, I laugh.

"What are you laughing at?" asks the palaeontologist.

"At the word *ominous*," I admit.

"Right," he says.

I like that he finds laughing at such a doomy word quite normal. One of the good things about Arsuaga is the way he accepts nonsense very naturally.

All of a sudden, we emerge from the mist as if emerging from a dream, and there before us are the Atapuerca mountains; then we turn away, leaving them to our left, to drive into San Millán de Juarros.

"We're in the Juarros region now," the palaeontologist tells me. "'Juarro' comes from the Basque for 'elm'; medieval Castilian used the same word. We've got a little further to go still. We're heading to Salgüero, which is famous for its international shearing competition."

"Shearing as in *shearing*?" I ask.

"What else is it going to be?"

"Like, as in shearing sheep?" I insist.

"Of course," he says.

"And what are we going to Salgüero for?"

"We're meeting some friends. You'll see."

Once we've arrived, he parks the car outside a kind of cantina that though in Burgos could just as well have been in Tijuana, and we go in for a coffee. It's a low-ceilinged place, very low in fact, like the cloud cover (the day is refusing to brighten up), which helps to keep the place warm, essential when temperatures outdoors are down to around freezing.

"It's going to rain," I say, coffee cup in hand, gesturing outside.

"No matter," says Arsuaga. "When it rains, you get wet. Then you dry off, and that's that."

"When I get wet, I catch a cold," I warn him.

On one of the walls of the establishment, I spot a poster that is part of a campaign against the threat to transform the region into a huge wind farm. It shows those electricity-producing wind turbines that are altering the landscape accompanied by a slogan that's actually kind of brilliant: "They're not windmills, they're giants."

"If this project goes ahead," says Arsuaga, "it would mean the disappearance of the Juarros region, and people here know it. Not long ago, gentlemen with briefcases started showing up, offering people however much a year to install a turbine on their land. The people were delighted, because it was like money for nothing. Now they've realised the existential danger that comes with it."

"And what's the alternative?"

"Well, quite. What's to be done? None of us likes turbines, or hydro, and obviously not nuclear either. So how do we get out of this mess?"

"I don't know — how do we?"

"Whatever the answer is, it absolutely must involve stopping thinking about science as though it were a religion. Once upon a time, the Church would solve the big problems of the day with rogations and novenas and so on. People now trust science in the exact same way. And that's because we aren't prepared to give anything up. We need to stop flying, stop consuming such ridiculous quantities of energy. But no, no, people say, anything but that. The solution has to be magical, somehow, and it's science that's got to give it to us. There's no way for us just to face the problems in a realistic way."

"Right."

"That said, I actually prefer not to talk about climate change, ever since a conference, twenty or thirty years back, in Tolosa — that's where my father's from. When I explained the problems we're facing, the other person on the panel said, 'There's a solution to this: Buddhism.' But it wasn't this that surprised me, so much as the enthusiastic ovation from the people in the hall that followed. I just thought, there's nothing to be done, it's literally hopeless."

While we're having our second coffee, next to the cantina exit, a tractor pulls up outside. The driver joins us, and the palaeontologist introduces him as Eduardo Cerdá, director of Living Palaeolithic, a private initiative that aims, as its representatives say, to "promote and preserve nature, endangered species, and their equilibrium with mankind, via the protection and study of ecosystems".

Briefly: Cerdá and his people have managed to reproduce — mere kilometres from Burgos — a couple of hundred hectares of Prehistory, which is home to European bison; Mongolian Przewalski's horses; the tarpan, a "reconstruction" of the wild European horse produced by crossing various breeds of domestic horse to which it was quite similar; the auroch, a "reconstruction" of the wild bull produced by crossing various cattle breeds; and cattle from the Scottish Highlands.

"Within two months," Cerdá tells me, "we'll be getting deer and reindeer. We're growing."

"It's a protected natural park?" I ask, thinking back to Cabañeros and the "as ifs" I experienced on my crossing from Rome to Majorca.

"Of course," he replies. "It's designed to bring Prehistory and its main characters closer to our visitors. I'm mostly in charge of the managing the tourism aspect, of coming up with workshops and the educational side. Estefanía Muro, who you're about to meet, is the biologist — she takes care of the animals and the ecosystem."

"So I'll get a chance to experience the Palaeolithic like it really was?"

"Yes," he says, "we're going to travel back fifteen thousand years."

In the short trip from the cantina to the park, I ask Cerdá if he's involved in the anti-wind-farm campaign.

"Yes!" he replies. "All of us here are against it, because Burgos is one of the provinces that's already suffered most in the whole Iberian Peninsula. Wind turbines are being put up, without any sort of controls, on the land of our forefathers. Their whole legacy is being sullied, destroyed. All the lovely green places have disappeared thanks to the introduction of those monsters. The latest-generation ones can be two hundred and eighty metres tall. The damage to the landscape is irreversible, and it leads to massive environmental devastation of birds of prey, among others. They crash into the blades and are sliced in two."

"How much do they pay per turbine?"

"Oof, I think it's six thousand euros or more per year, at least. If there are ten of them, you could take sixty grand, at least. There are some district councils that collect so much money they don't know what to do with it and just work on pruning the forest so the trees look pretty."

"The turbines are always on municipal land?"

"I'm not aware of any on private property. Everything they're planning here affects public land. All the villages are against it. They don't want more wind turbines. They don't want the money because, ultimately, in that landscape, who's going to want to come here? Who's going to deal with the subsidence in that house over there, who's going to build a country hotel? Who's going to come on family holidays here with all that around? How will a Living Palaeolithic survive in a place surrounded by wind turbines? It's not natural. Yes, it's precious clean energy, but Spain has plenty of coastline for putting the wind turbines out in the sea, like they do in England. On the coast, they wouldn't do so much damage. Besides, it's unfair that in places like the Basque country or Cantabria, they're hardly putting any up at all! We idiots in Castile get the wind turbines, our landscape gets ruined, but the people who benefit from the clean energy aren't us, it's the rest of the country."

When we get out of the car, after a very narrow dirt track with some incredible forest on both sides, the clouds part like two halves of a wound, and the sun, through this cleft, illuminates a vast meadow, with trees clustered here and there — a meadow that looks like the sort of prehistoric landscape we're used to seeing in encyclopaedias and movie documentaries.

"It's not just that it's cold here," says Arsuaga, zipping up his anorak, "it's that the cold here comes alive."

The cold is indeed intense and occasionally accompanied by a rain so fine one might mistake it for mist. The woodland we can see beyond the meadow is too much: I can't mentally digest it, nor tell myself anything original about it. I'm moved by the power of the ineffable, of what cannot be spoken, as if these meadows and these forests have awoken something that was asleep inside me and that has just been roused so spiritedly that it's a little scary, even though this is a fear that does also produce a strange euphoria. The damp air penetrates my nostrils, which seem to me to be prehistoric nostrils, and escapes from my mouth in the form of steam after having hydrated my skull.

That's how it feels to me, anyway.

"In this meadow," says Cerdá, "there were thousands of head of cattle until we joined the European Union and they all got slaughtered. Not a single cow was left, off they all went to the slaughterhouse. They paid a lot of money for every cow slaughtered. The meadow was left deserted. The cows used to fertilise this ground with their dung. I imagine the dung beetles wondering what had happened, why from one year to the next there wasn't a single cowpat left. Well, the dung has come back now."

We step into the forest. The ground is covered in oak leaves.

"Some of these oak trees," says Arsuaga, "are five hundred or six hundred years old. They're genuine sculptures, monuments. This hill is common land, and there was a time when it would be leased out to cattle farmers. When all that went, it was then that the Living Palaeolithic project came about, in connection with the Atapuerca dig."

After a short walk through the thick forest, trying to find

our way back into the relative light of that cloudy day, we emerge into a clearing where we see a group of horses.

"Are they prehistoric?" I ask.

"More or less," says Cerdá. "The wild horse called the tarpan became gradually extinct over the course of the nineteenth century. In reality, it became diluted with the domesticated breeds. This one's the Konik breed, from Poland, which retained a relatively primitive phenotype, which was then stripped back through selective breeding to recover the ancestral phenotype."

We walk towards the horses, which are grey and, strikingly to me, not very tall.

"That's the normal height for primitive horses," Arsuaga tells me. "They were all ponies. Alexander the Great would have ridden something like this. Taller horses are very recent, from the Middle Ages, when they were bred like that. On the frieze in the Parthenon, there are people on horseback whose feet nearly touch the ground."

"The cinnamon-coloured one," says Cerdá, "is a Przewalski. It's from Mongolia and it looks very much like the prehistoric ones. It's one of the jewels of this project, as there are only fifteen hundred individuals still alive on the whole planet. Most of them, around six hundred, are to be found in Mongolia."

"Living free?" I ask.

"No, there are fewer than four hundred in the wild. We managed to get ten specimens thanks to some agreements with an association in France. It's practically an extinct animal, because it cannot be tamed. It's inclined to panic, like zebras, and so nobody, apart from the very occasional horse-breaker, can ride it. That's why it was hunted for meat, because it couldn't be ridden."

While the rain is getting worse, I look at this Mongol

horse inserted into the strange landscape in which we find ourselves. The twenty-first century visiting Prehistory, or rather a reconstruction of Prehistory. An "as if" of Prehistory. But the Przewalskis' tendency to panic, that's real. It's part of what's left of Prehistory: panic, which they show when we look like we're about to approach.

"Are there males and females all mixed together?" I ask.

"Yes," says Cerdá, "the herds are all clearly demarcated from one another, and have their own clear hierarchies. It's a male with his females."

"This Przewalski," adds Arsuaga, referring to one that's standing somewhat apart, "is an exile."

"Przewalskis fight to the death," Cerdá explains. "The dominant male pushes out any young males that start to show an interest in the females, and they do it with huge amounts of violence. It's not like with other horses, where they fight a bit and one of them runs away. These can lose their lives, and the injuries they inflict on each other are terrible."

"They're fighting to be able to transmit their genes," says Arsuaga.

"So as to prevent the visitors — especially when they're children — from seeing animals with ugly injuries," Cerdá continues, "we take out this group of males and just leave the dominant one with his harem."

"For as long as he can handle it," I suggest.

"Of course, because at some point another will appear that's stronger than he was, he'll be replaced, and he'll die here."

The rain stops and the sun shows itself for a moment, once again, through a crack that opens in the sky and then quickly closes again, like a blink. The clouds, which are very black, look like they must weigh tons.

We advance across the damp grass, encountering more herds of horses that look like they've been taken from drawings in prehistoric caves.

"There are no predators here," I venture, recalling Cabañeros.

"We've seen wolves around," says Cerdá, "but they've never attacked the animals or us. It would have to be a very hungry pack."

"So do you have to slaughter some of the animals?"

"When there's an overpopulation of domesticated horses, who foal every year, yes, we do. They get sold to dealers or to firms that want them for meat products. We use that money to self-finance the project. We need to be sustainable because we don't get any funding, and selling young males is one of our sources of income," says Cerdá, taking an extremely narrow path between the oaks.

I note pre-emptively that we are moving further from the car, so that if it starts raining hard (there are spots here and there, seemingly at random), there would be no way to avoid getting soaked.

"It's like Sherwood Forest," says Arsuaga, who seems to be perked up by the ozone in the air.

We stop in front of an oak that is many-armed, like a vast asymmetrical candelabra, which simultaneously suggests a human figure in a threatening posture. If I saw that at night, I think, I'd die of terror. The leaves of the oaks, which make up a forest floor of considerable thickness, crunch beneath our boots; they snap like hands that have been cut off and piled up on the ground. I ask if workers gather up these dry leaves, and Cerdá says no, they don't touch a thing.

"Whatever falls," says Cerdá, "can just rot. It's important

there's a natural filter. The rotting of the leaves enriches the subsoil and acts as substrate."

Soon we arrive at a kind of magic circle made up of six or seven ancient oaks, in strange and gnarly shapes, with a lot of moss on the side more exposed to the wetness. We can hear the drumming of the rain on the branches, but the water gets lost somewhere along the way and never reaches our heads. All three of us have noses red from the cold, and we rub our hands to get the blood flowing. I want to suggest going back, but I don't dare, as I get the sense the others still have something they want to show me.

"This oak, which is so wide," says Cerdá, stopping, "that even three people combined couldn't get their arms around it, was undoubtedly here when Columbus set off for America. It's nearly six hundred years old and it's still got leaves."

When we reach the pasture, the rain has stopped again, but the cold is more noticeable than in the forest. Arsuaga comes over, takes me by the arm, and points to something at the other end of the meadow.

"What do you see?"

"A prehistoric bison!" I exclaim.

"Right," says Arsuaga. "There are bison here, another animal that was virtually extinct, like the Przewalski."

Though we are still at some distance from the animal, its outline is unmistakeable. It's been engraved in Spain's collective memory ever since we first looked in wonder at that hyperrealist bison at Altamira. We've all got a bison inside us, and so seeing it not on the inside, but as if projected by our imagination into the middle of a prehistoric landscape and as part of the unfolding of an unpleasant autumn day that's beginning to feel Palaeolithic to us too, unleashes, at least in me, a thrill that is only partly mine, as it's a thrill that reaches right down into my guts, riding over those of my ancestors, as if between the umbilical cord of the men and women who sprang from this pasture thousands of years ago and my own, there had never been any break, and as if the blood in their bodies was the same that was being pumped, unusually strongly now, by my own heart.

Immediately the body of the bison who has shown us his profile is joined by more of his fellows, whose outlines, silhouetted against the wet, grey day, which seems to be made of charcoal, are also unmistakeable.

"Now, this really is pure Prehistory," says Arsuaga. "These animals would have grazed land identical to this."

"It's a matter of returning the species to nature, but only where it's right for it to be living," says Cerdá. "There were bison here in the Pleistocene, which is why they adapt so well and reproduce, though the bison of that period were of a somewhat different species. Along with some other natural parks, we're part of a network of reproduction and conversation both of the Przewalskis and of bison. These animals," he adds, "have a dreadfully close blood relationship because they come from a few survivors that were in zoos. If they were all together and there was some outbreak of disease, they'd all die. Since there's so little genetic variation, what would kill one would kill them all. The only way to avoid this risk is to raise them in separate clusters, so that if an epidemic occurs among the bison here, it can't be transmitted to the ones in Poland, which is where the main group is."

"You're part of a global network of parks?" I ask, amazed.

"Of course. There's an institution that manages the whole species."

"Now," says Arsuaga, coming over to me, "maybe you'll start to understand why sex still exists, even though there's such a thing as parthenogenesis. In parthenogenesis, which is many animals' means of reproduction, there's no intervention from the male — indeed, there aren't any males. This is the case, for example, with certain species of gecko, which produce exact clones of the mother without any insemination having taken place. 'Clone' is a word that's used for an individual that's identical to its progenitor, but originally it was used to refer to a line, a set of individuals that are genetically the same. Now it's become fashionable, and is used more to refer to the individual. Clones perpetuate themselves over time without their genes being modified, and they can go on for thousands or millions of years."

"In that case," I ask, "why do we need sex, which gives us so much satisfaction but also creates so many problems?"

"When a female is parthenogenetic," replies Arsuaga, "her offspring have her genes and nothing else, whereas if she reproduces with a male, only half the offspring's genes come from her. To a Darwinian, for whom the individual's genetic interests are everything, this implies relinquishing half of our genes each time we reproduce. The other half comes from the mate."

"So what's the reason for this relinquishing?" I ask.

"If there's no genetic diversity," he says, "the genetic line always remains the same. But in the arms race against pathogens — protozoa and all the different bacteria and viruses — all those enemies are gradually evolving. If the species doesn't evolve, too, to defend itself against the new and improved pathogens, sooner or later all its members will perish."

"Right," I say, breathing in the ambient dampness, never

taking my eyes off the Palaeolithic bison, just in case they notice us and decide to charge.

"All living creatures," continues the palaeontologist, "are in a constant arms race with pathogens. We've seen it with Covid. If no vaccines had appeared, many more people in those parts of the population most susceptible to the illness would die."

"And those that survived," I deduce, "those who were young enough, would have reproduced and passed on their immunity to their descendants."

"And that's all there is to it," says Arsuaga. "It's exactly what happened with the Spanish flu: those who were susceptible to the H1N1 subtype of the virus died, and the species carried on. That was an end to it, more or less — from the point of view of the overall human species, of course, because it was a catastrophe at a societal level."

"In other words," I say, "sex is a great invention."

"An invention to tackle the stepping-up of pathogens' firepower," he insists. "When sex appeared, it produced genetic variety, i.e., there being no two individuals the same: everyone's unique. Twins aside, obviously, as they're genetically identical. If H1N1 were to have another major outbreak — and there were pandemics of it in 1977 and 2009, only then we called it 'Russian flu' and 'swine flu' — there would be some individuals who could withstand it and others who couldn't. Those who do withstand it will transmit their immunity to their offspring. But viruses will continue to mutate. And round it goes."

"And so sex," I venture, "favours the species more than the individual."

"Indeed, because the individual's interests would actually be better served by parthenogenesis. I've talked about pathogens, but what goes for pathogens also goes for the environment.

Imagine there's a change in the climate: if they are clones, they might all get wiped out, while if there's genetic variety, some will withstand it and others won't. In other words, they can evolve. The important part to get into our heads is that there's a direct relationship between the genetic variation of a species and the speed at which it evolves. Species with little genetic variation evolve more slowly than those with lots of genetic variation."

"And is there no contradiction between Darwinian theory, according to which natural selection only cares about the individual, and the appearance of sex, which favours the species? How does Darwinism explain that contradiction?"

"Because, effectively, there's a level of selection above that of the individual: group selection."

"And does Darwinism accept that?"

"What else can it do! It's the only possible explanation for the whole thing about sex."

"And so," I deduce, "sex exists not because it's convenient for individuals but because it suits the species. So *now* we know who it is we're fucking for!"

"Imagine these bison are clones and they never evolve. In fact, they almost are, given their close blood relationship. There is some sexual crossing, but no genetic diversity: the genes of the male are almost the same as those of the female. To all intents and purposes, we could say they are clones."

"Hey," I say, hoping to escape, "it's started raining again, we're going to get soaked."

"Don't worry, we'll dry off afterwards. Listen to this: all around them, around these bison, everything's changing. The wolves, for example, are getting faster, stronger, meaner; the climate's changing and, in place of oak trees, conifers are proliferating. And of course, the fungi, the protozoa, the bacteria, the parasites, and the viruses are gradually evolving to become more efficient at what they do, which is living and reproducing at the expense of others."

"But the bison are still the same," I jump in, to speed things up.

"In this arms race, they're becoming progressively weaker."

"Okay."

"It's called 'the Red Queen hypothesis', after the character in *Through the Looking-Glass*. At a certain point in the story, Alice and the Red Queen are standing together beside a tree. The Queen takes Alice by the hand and they start to run. They run and run, getting faster all the time. After a while, Alice starts to get breathless, and they stop to rest, but then she exclaims: 'But we're still in the same place, next to the same tree.' 'And what's strange about that?' asks the Queen. 'In our country,' says Alice, 'you generally get to somewhere else, when you run for a long time.' 'You must live in a slow sort of country,' says the Queen."

"This means," I say, "that from an evolutionary point of view, you've got to run a lot just to stay in the same place."

"Precisely. If everyone's running, and you're standing still, you're going to get eaten by all those who have sexual reproduction. There has to be variety in order that the best combinations can be selected. Stand still, and it's curtains."

"So you've been telling me this whole time that selection only cares about the individual, and actually it turns out it's also concerned with the species."

"No, no! It isn't that it favours the species. You're thinking in straight lines, and you'll never get anywhere that way. You have to think about it the other way around. Sex doesn't favour the species. That's teleology, magical thinking. Sex doesn't favour anyone. In nature there's no aim, no preferences — be a little Epicurean for once. It's all just atoms combining and separating."

"But that's what you just said, for fuck's sake!"

"I didn't say that."

"You did!"

"I said," he adds, smiling in the rain, "you need to make sure you keep running all the time if you want to stay in the same place."

"You said there's a kind of selection above that of the individual: group selection."

"Yes. Jacques Monod received his Nobel Prize jointly with André Lwoff and François Jacob. Jacob talks about there being three types of memory in animals: that of the nervous system (the experience that's built up in the neurons), that of the immune system (if you get infected by a pathogen and survive, you'll be left with an immune memory that's a kind of vaccine, since you've developed defences against the pathogen), and that of phylogenetic memory, which is evolutionary memory: the tree you see over there contains genetic memories of its predecessors; it comes from its modified predecessors. There's an important difference between these three types of memory. Everything that's stored in the nervous and immune systems allows you to learn from your mistakes: a cow gets a shock from an electric fence, and it knows not to go near it again. If you get some kind of flu, the immune system learns to defend itself against the virus. But evolutionary memory learns only from the successes, never from the mistakes."

"The mistake being death."

"Extinction. The genetics of the species are the result of predecessors who learned from the successes and only the successes. There are no second chances. But it isn't that sex shows up 'for' anything; you need to put the 'for' out of your mind altogether. There are sexual and asexual species. Full stop. What happens is that, among the asexual species, evolution occasionally deals them a bad hand. They appear and disappear because, sooner or later, there will be a change in the climate or a pathogen will manifest that they can't adapt to. It's not that sex exists in case a change in the climate happens some time down the road, in millions of years ... No. Evolution has no eyes for the future. A change in climate will come, or a pathogen will show up, and it's simply that species without genetic variation and sexual reproduction will disappear."

"Well I'm not sure I understand," I say, wiping the water from my face, resigned to the fact that I'm going to catch pneumonia, possibly to die from it. I'm consoled that this is the only chapter in the book left to be written and that the palaeontologist can finish writing it. It would be dynamite from a publishing point of view. I'm almost wishing for the icy rain to worsen, so I can occupy the vantage point of death to watch the success of a book that is itself about death.

"Okay," says Arsuaga, somewhat irritated, "if you want me to tell you there's a level of selection above the individual in geological time, then yes, there is — but this is *geological time*, which is measured, to give you an idea, in millions of years."

At this point, Estefanía Muro, the Living Palaeolithic biologist, appears through the rain. She tells us that the American bison, which also hasn't got any genetic variation because it too almost went extinct, suffers from an exceptionally rare viral illness that affects the penis, the glans, and that a lot of them needed to be slaughtered, so they're almost set to be wiped out.

"The only solution," says Arsuaga, "is to take away all the

ones that are ill. They're all very exposed because the viruses are constantly evolving, doing what they do, mutating, while the bison are still the same as ever — the Red Queen hypothesis."

"Actually," says Muro, "one of the ideas of this project, while we're waiting for some genetic diversity to appear, is the creation of different population centres. If we manage to arrange things, for example, in such a way that not all the bison in the world are in Poland, we'll make sure that, if the virus hits Poland, the bison there would die but we'd retain a reservoir of genes. Here with the Przewalskis, we've had problems with inbreeding: there used to be a dominant male whose young had tendons that didn't work. Some were stillborn, or only the head came out and both mother and baby died. We removed that individual, put in another, and for the moment we haven't had any problems with the horses."

Just then, the clouds part once more, and another ray of moribund light emerges, reminding me of the foal's head emerging from its mother's body. In spite of everything, the rain is growing steadily lighter, and so a rainbow soon appears. Nature, I think to myself, can get very rhetorical sometimes. Too many phenomena all at once to be able to appreciate them.

"This year," Muro continues, "we've removed this new male's firstborn daughter. She's four years old, and from that age they're fertile. To prevent the male from mounting his daughter, we've moved her to another reserve, in this instance the one in Palencia. We're going to have the same problem with the bison. We need to remove the firstborn female to stop her being mounted by her father."

I'm troubled by the idea of so many fathers trying to mount so many daughters. I'm troubled by inbreeding, I'm troubled by genes, sexual reproduction, parthenogenetic copies, rhetorical rainbows, I'm troubled by the rain, the sun, the cold. I'm troubled by the small group of prehistoric bison that are so exposed, poor beasts, to pathogens, all because of the Red Queen hypothesis. They're standing a long way away, but from time to time the dominant male, burdened with a hump that reminds me of the boulder of Sisyphus, turns his head and observes us as if wondering how four twenty-first-century people have suddenly dropped into the middle of the Stone Age. I'd like to explain to him that actually the opposite is true, they're four prehistoric animals who have travelled to the age of computers.

The biologist starts walking towards the group of bison, and the rest of us follow, amazed by the effect on our spirits produced by view of the forest from the pasture.

When we're close to the animals, I ask Muro if they're peaceable.

"It depends," she says. "They have their moments. But now they've already eaten, and there aren't any females on heat. Look, there's three bulls, and one of them's the alpha. The other two have very tough lives."

"In normal conditions, not in a protected park, but out in the wild, would those two have left the group?"

"They'd have formed their own herd, yes. I'm fighting to get one of them out of here because we can't get him to be well. He's in a state of constant anxiety."

I look at the anxious bison and it's not hard to recognise his distress, which is manifest in a physique that's very lacking, if you compare his solidity to the leader's.

"We've completely rid him of all parasites," says the biologist, "and we've given him vitamins, but he doesn't get any better."

"In nature," says Arsuaga, "he'd have left, but he'd be having a shitty time. Leaving home's not all that easy. It's cold when you're away from the herd. Normally, the alpha male is in the best spot, he's the one who gets the best grass to eat. So if you're alongside the alpha male, in spite of the stress, you do wind up in the best pastures. Leaving the group means having to make do with the worst spots in the territory."

"Does the dominant male do the *most* screwing, or is he the *only* one who gets to screw?"

"What matters is that we make sure he doesn't screw us," says Arsuaga. "How good are you at running?"

"Me? Not good."

Cerdá and Muro laugh.

"Let's move out of his way," the biologist suggests.

"And if we move out of his way, nothing's going to happen?" I ask.

"We'll soon find out," she says.

Curiously, looking at the profile of a bison's face, you can make out human features, identical to those you see on bison in the prehistoric caves.

"This female," says Muro, pointing to one of the nearer bison, "is very sociable. She's the bull's favourite, and she's the one in charge in the group."

It's stopped raining again, but the clouds have closed, the faint beams of sunshine have been cut off, the rainbow has vanished. The irregular ceiling of black clouds is moving closer to our heads, as if too heavy to stay up high.

Although we haven't covered the whole park, not even close, I manage to get them to agree to start heading back before the heavens really open on us. Everyone else is wearing more suitable clothes than I am.

As we head back to the car, Arsuaga says to me, "In the same way that sex has no explanation from the point of view of individual natural selection and can be understood only on a geological scale, there's a handful of biologists who claim that the same might go for death. Natural selection favours only success, and death is a failure — nobody wants to die ..."

"Inexplicable it may be," I say, "but it must make up part of our genetic information, because everybody does die."

"That's what proponents of the pre-programmed death idea say, that it's an adaptation on the part of the species. But I don't buy that, firstly because there isn't just a single cause of death. There doesn't seem to be any mechanism that programs death, given the fact death can have multiple causes. The best example is cancer, which is difficult to eliminate because there are many different kinds. If death were pre-programmed, we'd all die of the same thing. Besides which, the advantages of a pre-programmed death aren't at all clear."

"It leaves some space for the next people to come along," I say.

"Actually, the jury's still out on that. People say death is adaptive because it controls population size. But in Cabañeros, we saw that wasn't the case. The deer don't self-regulate. They just go on proliferating until they've eaten their way through everything. Didn't you see that documentary about Yellowstone, where the reintroduction of wolves led to a drastic reduction in the deer population, and in turn the decimated vegetation grew back?"

"But fruit flies, to give one example, get old in a way that's very similar to humans."

"True. And octopuses and Pacific salmon die very soon after reproducing, they do, but with all the signs of old age. They literally die of being old."

"So can we say that ageing is pre-programmed or can't we?"

"Well, there are arguments either way, and you've just put forward one of them. The Pachamama-ites need a bit of a chance," he says, ironically. "At any event, don't forget that the pre-programmed death hypothesis implies the existence of a genetic algorithm tasked with said programming, which would then simply need to be hacked in order to achieve eternal life. On the other hand, if there's no such thing as pre-programmed death, and each and every one of the illnesses we die from depends on a different genetic algorithm, they'd *all* need to be hacked. No mean feat. In short, the proponents of pre-programmed death are offering you eternal life in the way religion once did. We've changed provider, not the product."

We have a (good) meal at Los Claveles, a nearby restaurant that has been nurtured, like so many things, by the heat of the Atapuerca project; the family who run the restaurant have known the palaeontologist ever since he started working at the archaeological site forty years ago. The whole family comes over to say hello, and they recommend the best things on the menu.

I take a quick trip to the bathroom, where I dry my hair a bit with the hand-dryer and tidy myself up generally, since I've come straight from Prehistory, and am shattered from the trip, and people were giving me funny looks. While I'm at it, I give myself a quick respiratory check-up to ascertain the state of my bronchi, and, miraculously, all is well. I don't have the beginnings of pneumonia, or even the first symptoms of a common cold. Maybe Arsuaga was right: if you get wet, you get wet. Then you'll dry off again and that's that.

The food and the wine, in addition to the surprising state

of health in which I've returned from my cold and rainy experience, really lift my spirits, and I'm soon enthusiastically ready for the next activity of the day, which will take place at Burgos's Museum of Human Evolution, the only one in the world devoted to the human being, and whose contents Arsuaga is in charge of.

The first thing you notice about this museum is its container, designed by architect Juan Navarro Baldeweg and comprising a vast four-storey glass box whose parts evoke the different corners of the trench where the Atapuerca site is found, just a few kilometres away. The virtues of the building, devised expressly for the function it fulfils, are such that it's worth visiting for the intelligence of its spaces alone; the way it all interacts with the light is linked to the way we humans relate to the sun. It's a genuine pleasure losing yourself in its interior, in just the same way we lose ourselves, while falling asleep, inside ourselves — finding little corners of thought, of ancestral fears or distant emotions whose existence we can't even begin to fathom. For anybody with any curiosity about themselves or the species they belong to, it's unmissable. It's a museum you leave transformed, as if, without stopping being whoever you were when you went in, just by passing through, you've added to your identity the whole existential load of those who preceded us in the hard work of becoming men and women.

To a great extent, a visit to the museum constitutes an exercise in responsibility.

Arsuaga and I go straight to the part of the building where one encounters, in a strange circle formation, ten hyperrealistic latex sculptures of a range of individuals, who together tell the story of the different stages of evolution. The circle begins with

Lucy the Australopithecus and concludes with *Homo sapiens*, via *Homo habilis*, *Homo erectus*, et cetera, et cetera. The realism of the sculptures is so exceptional that if a visitor of flesh and bone were to stand very still next to one of them, they could be taken for a museum piece themselves. *Homo digitalis*, perhaps.

Standing in the middle of this group of our ancestors, as if we are just part of the family, the palaeontologist says, "Over these two years, we've talked about the possibilities of extending human life. I've brought you here so you can see everything humans have already done, and which we might therefore do again. We're going to see how. For now, though, consider this hominid, Lucy, who's an *Australopithecus afarensis*, a species that lived three and a quarter million years ago. We know they lived about as long as a chimpanzee, forty or fifty years fully in the wild. Humans, however, even around the time of Altamira, would live to seventy — with a fair wind."

"The leap is amazing," I say, keeping my voice down, as if the group of figures looking at us can also hear us.

"So that's something we've done," Arsuaga says. "We've extended our lives from forty-something to seventy or so. That is, from five decades to seven. What we need to establish is what went on between Lucy's time and that of the Altamira inhabitants, fourteen thousand years ago. Lucy, as you know, was bipedal. When she died, she would have been about twelve or thirteen, and was either pregnant or had just given birth. She wasn't fully mature yet — her bones weren't completely formed."

"How tall was she?"

"A little over a metre. She would have weighed in at around twenty-seven kilos. But although Lucy died very young, adults of her species would live up to forty-five, and some, whether through very good luck or very good adaptation, even got close to fifty. So how could an evolved *Australopithecus* have increased its longevity? The mechanism is the following: if, by whatever means, you manage to get a large group of individuals living to the age of fifty-five, this group would suddenly show up on natural selection's radar, natural selection gets to act on them, but they also have an edge over everybody else, for the simple reason that they have one child more, or maybe two more. I've met a female chimp living in the wild who was forty-five and was looking after a two-year-old. Living to fifty-five is a significant advantage, and it has repercussions for the next generation. Let's imagine that those who live to beyond fifty are generally possessed of better genes than those who don't. They're better adapted. There will then be a greater concentration of these individuals' genes in the following generation, because they'll have had one or two more children. And this gradually means more individuals in the species who survive to beyond fifty. This, my friend, is natural selection in action. Its radar isn't really anything but a kind of sieve."

"But a lot of people have got to live to that age," I say, "to fifty-five, for the expression of the unfavourable genes to be delayed. One or two long-lived individuals won't change anything, since their genes, even if they're amazing, get diluted in the ocean of genes of the species."

"Of course. Lots of individuals need to live to beyond fifty in order for the longevity to go from five decades to seven."

"Did being bipedal have any influence on the lengthening of people's lives?"

"No, Lucy was bipedal and she lived no longer than a chimp. It would be due to something else, not being bipedal. Here," he gestures at another sculpture, "we've got the South

African *Australopithecus* species. They're from a different place to Lucy, but their lifespan was the same as hers, and the same as a chimp's: around forty-five to fifty."

"The leap hasn't happened yet," I say.

"No, but not only are they bipedal, their hands, if you look, resemble ours quite closely. But their longevity's still the same."

"And what about that one?" I say, gesturing at the next sculpture, since we're looking at them from left to right, as though reading a text.

"That's *Homo habilis*," says Arsuaga. "He still looks like an *Australopithecus*, but his brain's slightly larger. He's the first to come up with any technology. We don't know a lot about him, but he's closer to the figures we've seen so far than he is to the subsequent ones."

"But still no great leap as far as longevity's concerned?"

"A bit, possibly. But here's where we get the qualitative change," says Arsuaga, pointing at the hyperrealist sculpture representing *Homo erectus*. "We're at about the two-million-years-ago mark. First off, *Homo erectus* is taller, which also has an effect on the brain. Compare him for size with the previous ones."

"Yes, I can see that," I say, trying not to catch *Homo erectus*'s eye, as he seems to be trying to tell me something.

"This hominid is more technological than *Homo habilis*, and he has a somewhat longer lifespan than a chimp."

"Okay," I say.

"And here you have *Homo sapiens*," he says, pointing at the last of the figures in the circle that surrounds the two *Homo digitalis*, represented by the palaeontologist and me. Our flesh is identical to that of the latex sculptures who are looking at us from a spatial distance of two or three metres, though from a temporal distance of thousands or millions of years. "The people who come to the museum," he adds, "can tell intuitively that there are two types of species in human evolution: the bipeds who, in spite of their upright posture, are closer to being chimps, and those who are like us, which is everything that happens from *Homo erectus* onwards. It's with *Homo erectus* that lifespans really start to increase, and there's this increase in height, too, and in brain size, as well as a technological wealth and a social complexity that wasn't there before."

"Okay. The point being?"

"The point being, it was millions of years ago that hominids started to live longer lives. Speaking in geological terms, we've managed an increase of two decades in barely two million years. One decade for every million years! We've gone from the longevity of a chimp to ours in the geological blink of an eye."

"What does that mean?"

"It means there are certain buttons that can be pressed to make it happen, because we've already done that."

"When you say buttons, are you referring to genes? It is a matter of activating certain genes?"

"Take special note of this: we have roughly the same number of genes as chimps do, about twenty thousand. The differences between the two species are found in only one per cent of the genes; the other ninety-nine per cent are identical. Not long ago, it was thought that the difference between a chimp's genes and ours was enormous — not to mention the difference between a fruit fly and a human being. We now know that this isn't the case, that even the differences between us and a fruit fly aren't that considerable. From which I deduce that the increase in human longevity doesn't depend so much on the number of genes that have mutated, as on their regulation."

"And what do we understand by 'regulation'?"

"Genes are like light switches. They can be on or off. The ones that are on, we say they're 'being expressed'. So there are on genes and off genes — genes that are and are not expressed. How to explain, then, that with such a minor genetic difference between chimps and us, we're so different in terms of anatomy and behaviour?"

"Is it because there are genes that have been expressed in them and not in us?"

"Rather it's a case of *when* in the maturation process the genes are expressed, and *where* — in which cells. This is the key to the differences between us and chimps, apart from the one per cent of genes that are different, obviously. Genes are a part of our DNA, of our genome, but they're only a very small portion, because the majority of our DNA doesn't get expressed. That is, it doesn't translate into proteins — which is what the genes do, they describe proteins."

"It's a kind of dark matter."

"We don't know what it's for, but it's there."

"Is that what some people call 'junk DNA'?"

"They used to call it that, not anymore. It was precisely our friends Jacques Monod and François Jacob who discovered that genes can be on or off. Until then, nobody had any idea. Maturation consists of activating certain genes here at one moment, there at another. It may be in this unexpressed DNA — what you've just called 'dark matter' — that the key to genetic regulation lies."

"In a nutshell," I suggest, "you don't need to alter very much in our DNA to increase longevity."

"Right now," Arsuaga explains, "we're a long way from being able to do that. Our knowledge of developmental genetics is still too rudimentary. But if we let ourselves dream, it isn't an aberration to think that life might be prolonged. As I say, we've already achieved that, and in a very short space of time. By altering a very small number of genes and their regulation, we've made the leap from *Australopithecus* to *Homo sapiens*, a leap that affects everything, longevity included. Which means it can be done."

"Eternity as utopia."

"I don't know if I'd go as far as to say eternity, but again, we've already done it, we've increased our longevity, and it hasn't taken us long. And without any magic. Pure molecular biology. If we one day come to a comprehensive understanding of human genetics, and gene expression is something that can be artificially controlled ..."

We return to Madrid that same night, with the Nissan Juke's windscreen-wipers working overtime, as it's bucketing down. When we reach the Somosierra pass, in the dark, with the lights from the lorries, the occasional lightning flash on the wet roads, and Arsuaga's knowledge to guide us — he's been doing this trip every week for years and in all weathers — he says, "Let's repeat an exercise that I've put to you on other occasions."

"Go on."

"Let's imagine that in the future, people manage to suppress many of the internal causes of death. The purely structural ones: your eye lens gets replaced, or your heart's removed and you're given a pig's heart instead, or a mechanical one. A treatment appears that can cure Alzheimer's, or prevent it. Your joints are replaced with prostheses that are actually an improvement on the originals. Cancer's solved by boosting the immune system. All these causes disappear from one day to the next. Let's include neurodegenerative disorders, which are real buggers because the nerve cells can only die, not divide: the ones that are lost are never regained. So, on the one hand, say we manage to solve everything that's mechanical, and on the other we ensure that your immune system, far from declining, becomes more efficient — without going too far, of course, because if the immune system's *too* strong, it could start to destroy the healthy cells of our own body. It's good to have police, but we don't want a police state. Finally, we work at a cellular level to eliminate oxidative stress and other problems, like that of telomere shortening — again being extremely careful in order to avoid producing tumours."

"So all that's left, finally," I say, "will be external causes of death."

"Okay," says Arsuaga, "but now we start taking those out of the equation as well. We take out hunger, cold, heat, thirst, parasites, pathogens. What's left? Violence. Let's say we manage to sort that out, too, that the UN establishes a new world order, not the Big Brother kind, but one that sees an end to all violence. It's utopian, but let's suppose for a moment that it all works out."

"So what are we left with then?" I ask.

"We're left with the Somosierra pass at midnight on a rainy night like this, with someone like me — old and tired — at the wheel. Someone whose eyesight is not the best, and whose reflexes aren't the greatest anymore, either. We're left with accidents."

"Well," I say, feeling a little scared, "you just focus on the road now, I've got plenty of material for this chapter already."

"Wait," he says seriously, while I listen to the water pummelling the Nissan's roof. "Just wait. I've learned a few lines from *Leo the African*, the novel by Amin Maalouf, to be able to recite them to you in a moment such as this. Make sure you get this down word for word."

"Okay."

"So, a Muslim man dies in North Africa, and the alim who's preparing to bury him says: *If death was not inevitable, man would have wasted his whole life attempting to avoid it. He would have risked nothing, attempted nothing, undertaken nothing, invented nothing, built nothing. Life would have been a perpetual convalescence. Yes, my brothers, let us thank God for having made us this gift of death, so that life is to have meaning; of night, that day is to have meaning; silence, that speech is to have meaning; illness, that health is to have meaning; war, that peace is to have meaning. Let us give thanks to Him for having given us weariness and pain, so that rest and joy are to have meaning. Let us give thanks to Him, Whose wisdom is infinite.*"

The palaeontologist recites this text that he's generously learned for me with such seriousness that we spend the rest of the journey in religious silence. Within that bubble of silence, I am thinking that if death was indeed avoidable, we wouldn't have risked travelling from Burgos to Madrid on a winter's night as unpleasant as this one. If death could be avoided merely by not going anywhere, nobody would ever leave the house for fear of a roof tile falling on their head or being struck by lightning.

Fortunately, Arsuaga drops me home safe and sound, and arrives safe and sound at his place, just as we have arrived safe and sound at the end of this chapter and of this book.

All the same, maybe because goodbyes always conjure up in

me an unbearable sense of loss, I phone him the next day.

"We haven't talked enough about altruism," I reproach him.

"Cooperation and altruism," he says. "Of course, my dear Kropotkin. And consciousness. The great problem in science is consciousness and awareness, the existence of myself and the other. When consciousness emerges, it's related to cooperation … But this, provided we can bear one more minute in each other's company, we'll have to leave for another book."

"A book about artificial intelligence?"

"A book about consciousness, intelligence, and cooperation. We'd cover all the bases then, artificial intelligence included."

"Okay," I say, just to have the last word.